STANDARD PERFECTION
POULTRY BOOK

THE RECOGNIZED STANDARD WORK ON POULTRY,
TURKEYS, DUCKS AND GEESE, CONTAINING A COMPLETE
DESCRIPTION OF ALL THE VARIETIES, WITH INSTRUCTIONS
AS TO THEIR DISEASES, BREEDING AND CARE. INCUBATORS,
BROODERS, ETC., FOR THE FARMER, FANCIER OR AMATEUR

BY
C. C. SHOEMAKER

Published by Left of Brain Books

ISBN 978-1-396-31982-2

First Edition

Table of Contents

INTRODUCTION.

WHAT MAKES SUCCESS.

There is no royal road to success, even in chicken raising. It's a simple, plain road, and all who would reach the desired object must travel the road. Because it is looked upon as of small account, the poultry business has been neglected on many farms and allowed to fall behind other branches of farm husbandry. As a writer in the *New York Sun* says:

"On the farm, poultry raising has not been kept to the front; has not, in fact, began to keep pace with agricultural progress. Other branches on the farm are yearly improved; new methods and implements have been introduced that have materially increased the yield of the various products. Poultry raising, however, is at a standstill. It is no new thing to hear complaints of 'worst kind of luck' with poultry 'ever experienced,' etc."

What is necessary to do to improve poultry raising on farms? It may be answered by saying, follow the methods of the successful poultrymen, or, as farmers delight to call them, "poultry cranks." How do these "cranks" manage their flocks? What is their secret of success? Wherein does their method differ from that in vogue on most farms? These questions can easily be answered, and will show quite a contrast with the farmer's way.

First—The poultryman erects a comfortable house on a dry and elevated site. The roosts must all be on one level. Windows are arranged so the fowls will not be obliged to roost in a draught between them.

Second—The stock must be sound, healthy, active, vigorous and of some pure breed. Crosses, they know, are not to be depended on for general results. Occasionally they make a cross when it is desired to increase size—when they may be breeding from the smaller breeds. This cross is generally to fill an order for broiler chicks, or something of the kind. The cross breeds are never bred from, or rarely ever allowed to live a whole season through.

Third—Feeding is brought to a real science. The poultryman feeds his flocks with some aim—there is no "guess" work about it. He knows what to

1

feed to induce egg-laying; he knows that to fatten and get ready for market quite a different ration is necessary. He has found out that growing chicks which are making flesh, bone, muscle and feathers at one and the same time, need a varied diet of the most nourishing foods. Sloppy, cold-water-mixed cornmeal he considers almost a poison, and especially so when it is made quite an exclusive diet.

Fourth—Cleanliness is a necessity. Neglect to observe cleanliness will quickly outdo all other work. Clean quarters are as necessary as food, and just as essential to health. Fowls cannot breathe vitiated atmosphere continually at night and remain long in health. Unclean surroundings debilitate and render the fowls more subject to diseases. Vermin results from filth and neglect to keep the poultry houses in proper condition.

Fifth—The stamina of the flock is the anchor rock of success in poultry raising. It is secured by carefully selecting the best of each brood. Any unusually promising cockerel or pullet is carefully looked after. There is a distinguishing mark made in the web of their feet. A record is kept of those marked, making it easy and sure to select them when wanted for breeders. Eggs are saved from these best hens for incubation. This careful method of selection can soon be made to build up a flock's uniformity and at the same time wonderfully increase utility.

POULTRY MAXIMS FOR MEMORY.

A brooder is preferable to a hen for raising chickens, as they can return to the brooder at will, and will not be dragged around unnecessarily when tired. Do not crowd, and give them plenty of fresh water at all times.

When your space is limited, be all the more careful to keep the quarters clean, especially if the chickens can not get out much. During the warm summer months it would be better to clean out well at least every other day.

Geese are more distinctively grazers than any other fowls, and will keep the grass eaten off, as close as sheep. Besides the value of their flesh for food, the feathers are an item of considerable profit and should pay keeping expenses.

Scientific analysis tells us there is as much nutriment in a new laid egg as in a four-ounce mutton chop. It is unwise, therefore, to neglect fowls and feed them nothing but screenings. The birds must do more than simply live.

The raiser of scrub chickens sells his surplus stock at from two to three dollars per dozen, while the breeder of thoroughbred fowls sells at from twelve to sixty dollars per dozen, and often with little trouble to make the sales. Which do you think the more profitable?

It is especially true of the poultry yard that whatever is worth doing is worth doing well. An intimate every day and every hour acquaintance with the fowls is what leads to the profits. Do not become disgusted with their appearance at moulting time, but give them all the more care.

A Massachusetts farmer is reported to be making $4,000 a year out of his poultry, because he has caught on to the knack of doing things in the right way. The farm is the right place for chickens, and men or women with the right qualifications can make money if they make the effort.

A beginner in the poultry business should start with a few fowls, and gradually enlarge as he learns the requirements of the business. If one begins with a large number, he is liable to bring roup, lice, cholera and other undesirable things into his yards, which will bring disaster.

To make good layers, hens must have a regular and sufficient supply of egg-forming material. While they have free range in summer they can generally find this for themselves, but when shut up in winter they can not be expected to do well unless their feeding is well looked after.

Small waste potatoes, boiled or steamed till soft; form a cheap and useful occasional food; but the fowls soon become tired of them, and they should only be used at intervals.

Water fowls, ducks, geese, etc., do not require large bodies of water, as has generally been supposed. Some of the most successful duck raisers provide no water save for drinking.

Turkeys require and must have considerable range, and they must also be allowed full liberty with their young after they begin to feather, else they will surely not be kept in good health.

One of the secrets of success in poultry raising is loving the work. It should be a pleasure to take care of fowls.

If your chicks are not doing well, examine them closely and see if they are not infected with vermin.

Very fat fowls are poor breeders, and are liable to lay soft shelled eggs. Always avoid having your breeders fat.

To get eggs, avoid overfeeding, but do not starve. Green cut bone, scalded bran, oats and barley are good feeds. Give plenty of milk, if you have it.

PRACTICAL POULTRY.

CHAPTER I.

RAISING POULTRY FOR PROFIT.

Few of us realize how important a business poultry-raising is. It considerably exceeds in total market value the whole output of coal, iron and mineral oil in the United States. As a usual thing the poultry business is underestimated rather than overestimated. Thousands of people raise chickens merely to supply their own tables. They keep no record of the number of eggs they get or the chickens they kill. So this item in the total is neglected altogether, or is greatly underestimated.

The United States Census for 1890 gives the following figures for "farms only." To complete this summary, the poultry products of towns and villages should be added:

Chickens	258,871,125
Turkeys	10,758,060
Geese	8,440,175
Ducks	7,544,080
Dozens of eggs	819,722,916

The value of these products, estimating chickens as worth 40 cents each, turkeys and geese as worth 60 cents each, ducks at 45 cents each, and eggs at 15 cents a dozen, amounts to $241,418,660.

SMALL CAPITAL REQUIRED.

The poultry business is peculiar in that a small capital will start a person, and it is a great deal better to start in a small way and gradually rise to larger things as one gets experience and comes to know just what the market is and out of what kind of fowls one can make the most money. In no other kind of stock raising can a good start be made with so small an outlay at the beginning, since full blooded stock of other kinds is very expensive.

An ordinary farm is suitable for any poultry experiment, and no expensive tools or machinery are required in the conduct of the business. In other lines the expenditure for tools is large at first, and when the business increases, the old machines must be discarded for larger and better ones in order to produce economically. With poultry, this is in no sense true. The first cheap coops may be made over into better ones, and as the business grows the equipment is added to gradually and often out of the first profits of the business.

SUCCESS DUE TO SKILL.

Success in poultry raising is due largely to the skill and care of the raiser—more so than in almost any other line of farm production. Failure will almost inevitably be the lot of the inexperienced and careless person. The business must be learned, and learned thoroughly, and a person must have a love for it, if it is to prove a paying venture. So we advise the novice to begin in a small way, so that he cannot make costly mistakes for he is sure to have some trouble.

HOW MUCH LAND IS NEEDED.

Very little land is really required, unless you intend to raise all the grain the fowls will eat. If you can do that, of course, it will add to your profits. Chickens, even laying hens, will do well in a small yard if it is kept clean by cultivation; but turkeys demand more room to rove about in. Five acres will be sufficient for 800 hens, and allow space to raise the necessary green stuff. If the room is limited, however, the space should be divided up and the hens allowed to run only in small flocks as closely graded in size and vitality as possible.

A larger range is, of course, often useful when it can be had. Growing fowls require a great deal of exercise, and so do laying hens, especially young ones. The young fowls should have the largest possible run; old hens, which are laying, need less room, while fowls that are being fattened are not injured by close confinement—indeed they do better in confinement, if the confinement period of inactivity is not continued too long.

POULTRY GIVES QUICK RETURNS.

One remarkable fact about the poultry business is that it gives relatively quick returns. Years are required for cows, sheep and horses to grow, and pigs require a considerable length of time; but even if poultry-raising is begun by selecting eggs for hatching, the product in the form of broilers, capons or mature fowls should be ready for the market in from five to eight months.

Again, poultry raising is a business in which women may engage successfully; and many men who are not strong enough for more active farm work, or who need outdoor employment, after long confinement in the city, will find poultry raising a superior occupation. Even persons of leisure often find this a pleasant and healthful employment, which yields a large return in something beside dollars.

CHOOSE YOUR SPECIALTY.

Success in poultry raising depends to a considerable extent on the selection of the proper branch, for it offers a variety of specialties. Persons about to engage in the business should study well their own tastes and capabilities, and also their situation and means. Some will choose egg production, others will make a specialty of raising broilers, and still others will choose the raising of fancy fowls, or breeding. Thus if the space is limited, so that only a few hundred fowls can be kept, it will be advisable to confine oneself to egg-production (especially during those months when eggs are high in price), or to breeding fowls which will command a high price. Thus a profitable business may be done in the small space available. So, too, the various kinds of fowls, such as chickens, ducks, geese, turkeys, pigeons, etc., allow the poultry raiser a selection which he should make with care and reflection.

POULTRY EASY TO MARKET.

Poultry products are among the easiest to market of all farm produce. Eggs may be shipped hundreds of miles by express, and yet be indistinguishable from fresh ones, when a week later they are served as food; and meat fowls may be sent by freight or express for long distances. Of course, the farther they must be sent, the higher should be their grade, if the transportation charges are not to more than offset the profits.

It may be said that in the poultry business the best products are always more saleable and more profitable than the poorer and cheaper varieties. The demand for the best is strong and steady, while that for second grades is unreliable in the extreme. It is, therefore, manifest that skill and good work are the chief sources of profit in the poultry business.

CHAPTER II.

UNCLE SAM'S ADVICE TO POULTRY RAISERS.

Extracts from the United States Department of Agriculture's farmers' Bulletin No. 41:

The wide distribution of domestic fowls throughout the United States, and the general use made of their products make poultry of interest to a large number of people. Breeders are continually striving to improve the fowls for some particular purpose; and to excel all predecessors in producing just what the market demands for beauty or utility; but the mass of people look at the poultry products solely as supplying the necessary elements of food in an economical and palatable form. For a considerable time each year eggs are sought instead of meat by people of moderate means, because at the market price eggs are a cheaper food than the various kinds of fresh meat.

Large numbers of the rural population live more or less isolated, and find it inconvenient, if not impossible, to supply fresh meat daily for the table aside from that slaughtered on the farm; and of all live stock, poultry furnishes the most convenient means of supplying an excellent quality of food in suitable quantities. This is particularly true during the hot summer months, when fresh meat will keep only a short time with the conveniences usually at the farmer's command.

The general consumption of poultry and poultry products by nearly all classes of people furnishes home markets in almost every city and town in the United States at prices which are remunerative if good judgment is exercised in the management of the business.

Although fowls require as wholesome food as any class of live stock, they can be fed perhaps more than any other kind of animals on unmerchantable seeds and grains that would otherwise be wholly or partially lost. These seeds often contain various weed seeds, broken and undeveloped kernels, and thus furnish a variety of food which is always advantageous in profitable stock feeding. There is less danger of injury to poultry from these refuse seeds than is the case with any other kind of animals.

SELECTIONS OF SITES
FOR BUILDINGS AND YARDS.

Frequently the other buildings are located first and the poultry house then placed on the most convenient space, when it should have received consideration before the larger buildings were all located.

As poultry keeping is wholly a business of details, the economy of labor in performing the necessary work is of great importance. Buildings not conveniently located and arranged become expensive on account of unnecessary labor.

Visits and operations must be performed frequently, so that any little inconvenience in the arrangement of the buildings will cause not only extra expense in the care, but in many cases a greater or less neglect of operations that ought to be performed carefully each day.

To exclude rats and mice it is generally best to locate the poultry house at some distance from other farm buildings, especially if grain is kept in the latter. Provide cement walls and have the foundation below the frost line. Convenience of access and freedom from vermin are desirable points, and depend largely upon the location. Everything considered, it is safest to have the house quite isolated.

A dry, porous soil is always to be preferred as a site for buildings and yards. Cleanliness and freedom from moisture must be secured if the greatest success is to be attained. Without doubt, filth and moisture are the causes, either directly or indirectly, of the majority of poultry diseases, and form the stumbling block which brings discouragement and failure to so many amateurs. It must not be inferred that poultry cannot be successfully reared and profitably kept on heavy soils, for abundant proof to the contrary is readily furnished by successful poultrymen. The necessity for cleanliness, however, is not disputed by those who have had extended experience. When the fowls are confined in buildings and yards, that part of the yard nearest the buildings will become more or less filthy from the droppings and continual tramping to which it is subjected. A heavy or clayey soil not only retains all the manure on the surface, but by retarding percolation at times of frequent showers aids materially in giving to the whole surface a complete coating of filth. If a knoll or ridge can be selected where natural drainage is perfect, the ideal condition will be nearly approached. Where natural favorable conditions as to drainage do not exist, thorough under drainage will go a long way toward making the necessary amends to insure success.

CONSTRUCTION OF HOUSES.

The material to be used in the construction and the manner of building will necessarily be governed largely by the climatic conditions.

In general, it may be said that the house should provide warm, dry, well-lighted, and well-ventilated quarters for the fowls.

These requirements are a good roof with side walls more or less impervious to moisture and cold, suitable arrangements for lighting and ventilating, and some means for excluding the moisture from beneath. Cheap, efficient walls may be made of small field stone in the following manner: Dig trenches for the walls below the frost line; drive two rows of stakes in the trenches, one row at each side of the trench, and board inside of the stakes. The boards simply hold the stones and cement in place until the cement hardens. Rough and uneven boards will answer every purpose except for the top ones, which should have the upper edge straight and be placed level to determine the top of the wall. The top of the wall can be smoothed off with a trowel or ditching spade and left until the cement becomes hard, when it will be ready for the building.

The boards at the side may be removed, if desirable, at any time after the cement becomes hard.

For the colder latitudes, a house with a hollow or double side walls is to be preferred on many accounts, although a solid wall may prove quite satisfactory, particularly if the building is in the hands of a skilled poultryman. Imperfect buildings and appliances, when under the management of skilled and experienced men, are not the hindrances that they would be to an amateur.

A cheap, efficient house for latitudes south of New York, may be made of two thicknesses of rough inch lumber for the side and end walls. This siding should be put on vertically, with a good quality of tarred building paper between. In constructing a building of this kind, it is usually best to nail on the inner layer of boards first; then put on the outside of this layer the building paper in such a manner that the whole surface is covered. Where the edges of the paper meet, a liberal lap should be given, the object being to prevent as far as possible drafts of air in severe weather. Nail the second thickness of boards

11

on the building paper so as to break joints in the two boardings. In selecting lumber for siding, it is best to choose boards of a uniform width to facilitate the breaking of joints.

In constructing a roof for a house in the colder latitudes, one of two courses must be pursued, either to ceil the inside with some material to exclude drafts or to place the roof boards close together and cover thoroughly with tarred paper before shingling. The fowls will endure severe weather without suffering from frosted combs or wattles if there are no drafts of air. Hens will lay well during the winter months if the houses are warm enough so that the single comb varieties do not suffer from frost bite. Whenever the combs or wattles are frozen, the loss in decreased egg production cannot be other than serious.

Figure 1 represents a cheap and efficient method of building a poultry house with a hollow side wall. The sill may be a 2 by 6 or 2 by 8 scantling, laid flat on the wall or foundation; a 2 by 2 strip is nailed at the outer edge to give the size of the space between the boards which constitute the side walls. A 2 by 3 scantling set edgewise forms the plate, and to this the boards of the side walls are nailed. These boards may be of rough lumber if economy in building is desired. If so, the inner boarding should be nailed on first and covered with tarred building paper on the side that will come within the hollow wall when the building is completed. This building paper is to be held in place with laths or strips of thin boards. If only small nails or tacks are used, the paper will tear around the nail heads when damp and will not stay in place.

The cracks between the boards of the outside boarding may be covered with inexpensive battens if they are nailed at frequent intervals with small nails. Ordinary building lath will answer this purpose admirably, and will last many years, although they are not so durable as heavier and more expensive strips. The tarred paper on the inside boarding and the battens on the outside make two walls each impervious to wind, with an air space between them.

In preparing plans for a building, one of the first questions to be decided upon is the size and form of the house. If the buildings are made with the corners right angles, there is no form so economical as a square building. This form will inclose more square feet of floor space for a given amount of lumber than any other, but for some reason a square building is not so well adapted for fowls as one that is much longer than wide. It is essential to have the different pens or divisions in the house so arranged that each one will receive as much sunlight as possible, and to secure this, some sacrifice in economy of building must be made.

The writer prefers a building one story high, and not less than 10 nor more than 14 feet wide, and as long as circumstances require. In most cases a building from 30 to 60 feet long meets all requirements. If this does not give room enough, it is better to construct other buildings than to extend one building for more than 60 feet. It must be remembered that each pen in the building should have a separate yard or run, and that a pen should not be made to accommodate more than 50 fowls, or, better, 30 to 40.

The building should extend nearly east and west in order that as much sunshine as possible may be admitted through windows on the south side. The windows should not be large nor more than one to every 8 or 10 feet in length for a house 12 feet wide, and about 17 inches from the floor, or at such height that as much sunshine as possible will be thrown on the floor. The size and form of the windows will determine quite largely their location. In all poultry houses in cold latitudes the windows should be placed in such a position that they will give the most sunshine on the floor during the severe winter months. One of the common mistakes is in putting in too many windows. While a building that admits plenty of sunlight in the winter time is desirable, a cold one is equally undesirable, and windows are a source of radiation at night unless shutters or curtains are provided. Sliding windows are preferred on many accounts. They can be partially opened for ventilation on warm days. The base or rail on which the window slides should be made of several pieces fastened an inch or so apart, through which openings the dirt which is sure to accumulate in poultry houses may drop and insure free movement of the window.

VENTILATION.

Some means of ventilating the building should be provided. A ventilator that can be opened and closed at the will of the attendant will give good results if given proper attention, and without attention no ventilator will give the best results. Ventilators are not needed in severe cold weather, but during the first warm days of early spring, and whenever the temperature rises above freezing during the winter months some ventilation should be provided. Houses with single walls will become quite frosty on the inside during severe weather, which will cause considerable dampness whenever the temperature rises sufficiently to thaw out all the frost of the side walls and roof. At this time a ventilator is most needed, arranged with cords, so as to be easily operated. Figure 2 represents an efficient and easily operated ventilator.

PERCHES.

Perches should not be more than 2½ feet from the floor, and should all be of the same height. Many fowls prefer to perch as far above the ground as possible, in order, without doubt, to be more secure from their natural enemies; but when fowls are protected artificially from skunks, minks, foxes, etc., a low perch is just as safe and a great deal better for the heavy-bodied fowls. Convenient walks or ladders can be constructed which will enable the large fowls to approach the perches without great effort, but there are always times when even the most clumsy fowls will attempt to fly from the perch to the floor and come down with a heavy thud, which is often injurious, when the perches are too high.

There is no reason why all perches should not be placed near the floor. Movable perches are to be preferred. A 2 by 3 scantling set edgewise, with the upper corners rounded, answers every purpose and makes a satisfactory perch. The perches should be firm and not tip or rock.

Underneath the perches should always be placed a smooth platform to catch the droppings. This is necessary for two reasons: the droppings are valuable for fertilizing purposes, and ought not to be mixed with the litter on the floor; then, too, if the droppings are kept separate and in a convenient place to remove, it is much easier to keep the house clean than when they are allowed to become more or less scattered by the tramping and scratching of fowls. The droppings should be removed every day.

NESTS.

In constructing nest boxes, three points should be kept constantly in mind: (1) The box should be of such a nature that it can be readily cleaned and thoroughly disinfected; if it is removable so that it can be taken out of doors, so much the better; (2) it should be placed in the dark, or where there is only just sufficient light for the fowl to distinguish the nest and nest egg; (3) there should be plenty of room on two or three sides of the nest. It is a well-known fact that some hens in seeking a nest will always drive off other hens, no matter how many vacant nests may be available. If the nest is so arranged that it can be approached only from one side, when one hen is driving another from the nest there is likely to be more or less of a combat, the result of which is often a broken egg. This, perhaps, more than any one thing, leads to the vice of egg-eating. To the writer's knowledge, the habit of egg eating is not contracted where the nests are arranged in the dark and open on two or three sides.

DRINKING FOUNTAINS.

One of the difficult problems for the poultryman to solve is how to easily provide pure, fresh water for his fowls. Many patent fountains which are on the market are automatic and keep before the fowls a certain quantity of water.

A simple, wholesome arrangement may be made as follows: Place an ordinary milk pan on a block or shallow box, the top of which shall be 4 or 5 inches from the floor. The water or milk to be drunk by the fowl is to be placed in this pan. Over the pan is placed a board cover supported on pieces of lath about 8 inches long, nailed to the cover so that they are about 2 inches apart, the lower ends resting upon the box which forms the support of the pan. In order to drink from the pan it will be necessary for the fowls to insert their heads between the strips of lath. The cover over the pan and the strips of lath at the side prevents the fowl from fouling the water in any manner, except in the act of drinking. Where drinking pans of this kind are used, it is very easy to cleanse and scald them with hot water when occasion demands. See Fig. 3.

DUST BOXES.

It is necessary to provide dust boxes for the fowls during the winter months if they are to be kept free from lice. If the soil in the yards is naturally dry and porous, abundant opportunities will be had for dust baths during the warm summer months, but during the late fall, winter and early spring some artificial provision must be made. A comparatively small box will answer the purpose if the attendant is willing to give a little attention to it each day. These boxes should be placed so that they will receive some sunshine on each bright day, and be kept well filled with loose, fine earth. Road dust procured during the hot, dry months of July and August from much-traveled roads has no superior for this purpose. Probably there is no way in which the poultryman can better combat the body louse than by providing dust boxes for his fowls.

YARDS OR PARKS.

Where fowls are kept in confinement, it will be found best to provide outdoor runs or yards for them during the summer months. Give them free access to these yards whenever the weather will permit. The most economical form, everything considered, for a poultry yard, is one much longer than wide. Two rods wide and 8 rods long is sufficient for 50 fowls. Whenever a poultry plant of considerable size is to be established, it will be found most economical to arrange the yards side by side, with one end at the poultry house. The fences which inclose these yards may be made of poultry netting or pickets, and should be at least 7 feet high. In either case it is best to have a board at the bottom, for sometimes it will be desirable to give quite young chickens the run of these yards. If the poultry yards are constructed as described, there is sufficient room for a row of fruit trees down the center of the yard, and still leave ample room for horse cultivation on either side, either with one or with two horses.

These yards are to be kept cultivated. If thought best, grain may be sown before cultivation to furnish part of the green food for the fowls.

CHAPTER III.

POULTRY HOUSES AND BROOD COOPS.

The first thing to be done in establishing a poultry department on a farm is to select a location, and this is not a trifling matter, by any means, as a great deal depends on whether or not the selection is made judiciously and for the best interests of the fowls.

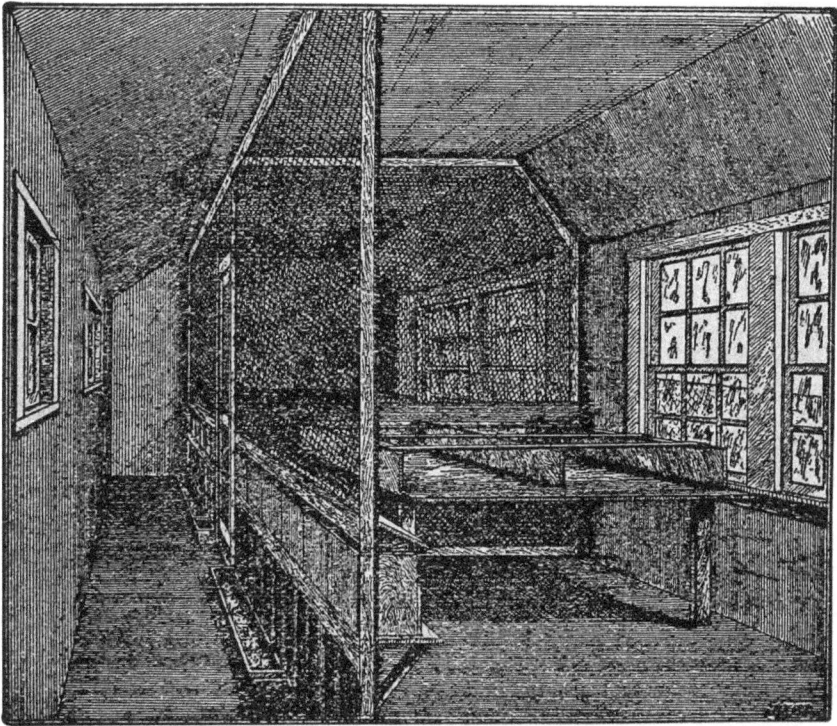

Interior of Poultry House No. 1.

A good compass location would be on the side of some hill that slopes to the south, but as this is not always convenient, the cold winds of winter can be kept back or partly broken, at least, by growing a foliage plant of some kind on the north side of the coop and yards. Small trees are the best protection but take considerable time to grow.

Poultry House No. 1.

Bear in mind there is *no best plan* for a poultry house. The best plan is the kind *you prefer*. Much depends on the cost, the location, the climate, the breed and the number of fowls. No two persons can well agree on the best plan of a dwelling house, nor is it possible to offer a plan of a poultry house that would be accepted by all as the *best*. A certain plan may be the best plan for you but not for your *neighbor*. Knowing these difficulties we have endeavored to give a plan which can be adapted to most locations and can be built of cheap material, or more substantially, as preferred. The ground plan of house No. 1 can be extended to any length desired with the same interior arrangement.

The dimensions of the house No. 1 are 14x24 feet. In the ground plan the alleyway at the north side of the building is 4 feet wide, and in the houses of greater length should extend the entire length of the house. F is the feed trough; placed in the alley to prevent the fowls from getting into it and for convenience in feeding; N is the nest boxes which are placed on a shelf 18 inches from the floor and arranged to open into the alley so that the eggs may be gathered without entering the pens. The cover over the nests should be placed at such an angle that the fowls cannot roost on them. The space under the nest boxes is lathed up, leaving space enough between the lath for fowls to feed through. The partitions between pens and over nest boxes may be made

18

of lath or wire netting. R is the roost which is placed 1 foot above a tight platform, the platform being raised 2 feet from the floor of the house. D B is the dust box. D is the door leading in from the outside, also from the alleyway to poultry department. Water may be placed in dish at end of feed trough. The floor may be either earth or boards, as preferred.

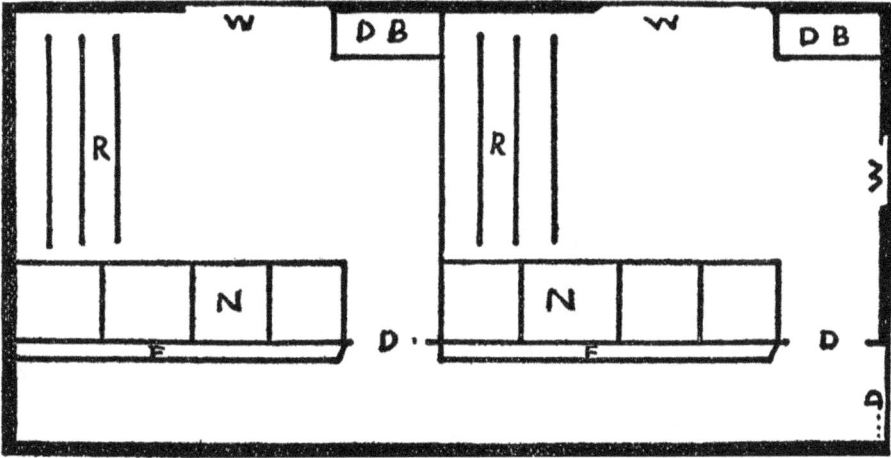

Ground plan of house No. 1.

POULTRY HOUSE NO. 2.

Poultry house No. 2.

The dimensions of ground plan of No. 2 are calculated to be 14x28 with 6 foot alleyway running crosswise between the two departments, and it will accommodate 25 to 35 fowls (according to the size of them), in each department and should be 7 feet high to the square. May be either made with shed or cone roof, but cone roof is preferable.

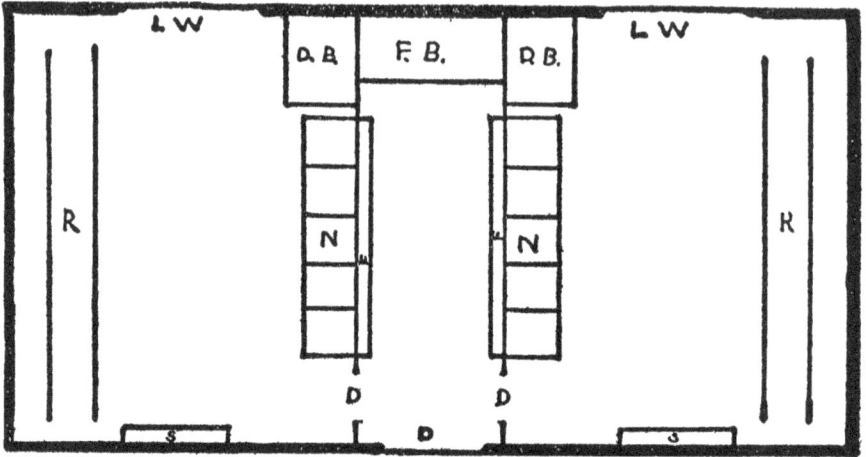

Ground plan of house No. 2.

Interior view of poultry house No. 2.

The building should face with windows toward the south. It may be built either with drop siding or stock boards, single wall or double, as preferred by the builder. R roosts are placed at far side of each department, 10 inches above platform, it being about 2 feet from the floor. The roosts may be cleated together on cross bars hinged fast to the wall, so when cleaning out the droppings you can raise them up and fasten them to the wall by means of a small hook attached to them for that purpose. S is a small box containing shell or grit for the fowls. F B, feed bin. L W is large window on south side; D B is dust box. D D D are doors leading from outside into the two compartments, N N are for nests placed 20 inches above the floor on a platform and are made to pull out same as a drawer, to gather the eggs, and may also be inverted so that while you are having a hen hatch she can come off the nest to eat, in the alleyway, and no others can get on the nest and disturb, or prevent her from going back. Nest boxes should be covered with a roof set at such an angle as to prevent fowls from sitting on them. The space from nest platform to the ground (20 inches) should be lathed up so the fowls can feed through from F trough placed in alleyway. Water dish may be placed at end of trough.

POULTRY HOUSE NO. 3.

Fig. 3 illustrates a single house with shed suitable for a small flock, which, if 10x16 feet, having an 8x10 feet shed and closed room of same dimensions, will comfortably accommodate 20 to 40 fowls, according to variety, the larger fowls of course requiring more space to a given number than the smaller ones.

Poultry house No. 3.

Those who have never given this method of winter housing a trial, should fit up at least one small house of this plan and give it a fair test the coming

21

winter. We are confident that the results in health of the flock, number and fertility of the eggs laid will be all that could be desired and that such a test will result in the adoption of some such arrangement for giving the entire flock the benefit of this 'open air fund" in the future.

POULTRY HOUSE NO. 4.

A house built with the arrangements of open sheds connected with a closed room in which the fowls may roost, lay or remain in very severe weather, is held in great favor by those who recognize the benefit and importance of open air exercise to the flock during winter. This arrangement is illustrated in Fig. 4, which shows a succession of closed room and open shed, which may be extended to any length desired when the site is sufficiently level to allow it. Each closed room opens into its respective shed, affording the fowls their choice between the two rooms, and it will be found that they much prefer the open air even on quite cold days in winter, especially if the shed floors are kept dry and well littered with straw, chaff or leaves, etc. If the grain feed is scattered in this litter the fowls will busy themselves the entire day in searching for it, and this open air exercise will prove an important factor in keeping the flock in health and inducing egg production. The roof and north wall of these sheds may be of the same material and construction as the house; merely omitting the south walls of the building at each "shed space"; a dry dirt floor will be suitable for these sheds if the same is filled so as to raise it eight or ten inches above the surrounding level, but for the closed rooms we would prefer a board floor, as these rooms should be kept very clean, and a board floor is more easily swept than any other.

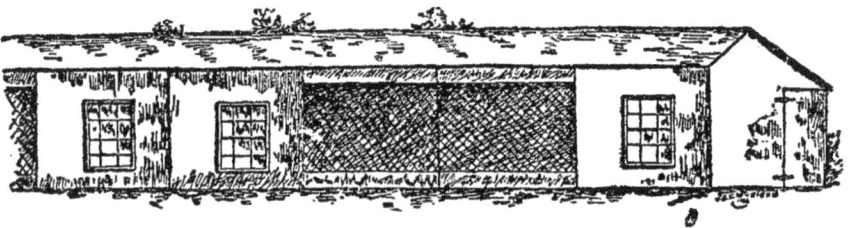

Poultry house No. 4.

In Fig. 4 the sheds are shown with fronts enclosed with wire netting to prevent the fowls of the different pens from getting together; a curtain could also be provided, to let down over the open side of the shed during extreme cold and cloudy weather and thus give the fowls the use of their "exercising room" without undue exposure.

Poultry Houses Nos. 3 and 4 are by courtesy of *Poultry Tribune*, Freeport, Ill.

POULTRY HOUSE NO. 5.

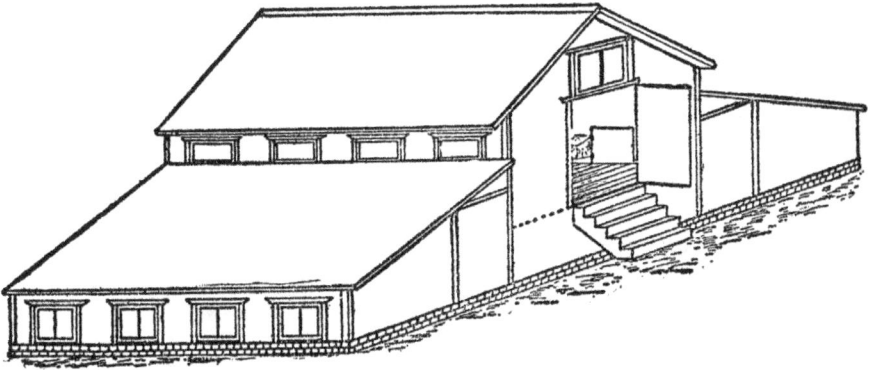

Poultry house No. 5.

This house is 20x32 feet, sill measure. Center part 12x20 feet, 9 feet to the square, having a solid floor 3 feet above top of sill (dotted line shows where floor should be). This part of the house is the roosting and laying compartment. Ground plan shows partition through length of building, made of plastering lath, and flooring up and down in roosting room. The frame is of studding 2x4 inches, weather boarded with worked barn boards, also lined with same, with smooth side out, having building paper tacked to studding under it. Roosts hang from ceiling on iron rods to within 14 inches from floor; a board set in grooves in wall divides floor space under roosts to keep droppings from mixing with litter on balance of floor. Feed troughs are placed around wall as shown in plan. Ladders lead up from scratching floor to each roost room. This roosting compartment is sealed up on under side of rafters, which permits of a window in each gable above the square of the building, giving splendid ventilation without any patent arrangements. Each window is hinged to bottom of frame and can be opened or closed to suit the weather. Wings are 3 feet high in front and 7 feet where they join roost rooms. Nest boxes can be placed where convenient as it is always best to use some kind of portable ones. Sheathing of hemlock, covered with building paper, then shingled with hemlock shingles; cost of material above foundation, $89.42. Foundation can be made to suit any one's fancy, but should be made six or eight inches above natural lay of land, and filled in to bottom of sill with gravel or dry soil of any kind, or cut corn fodder for scratching material.

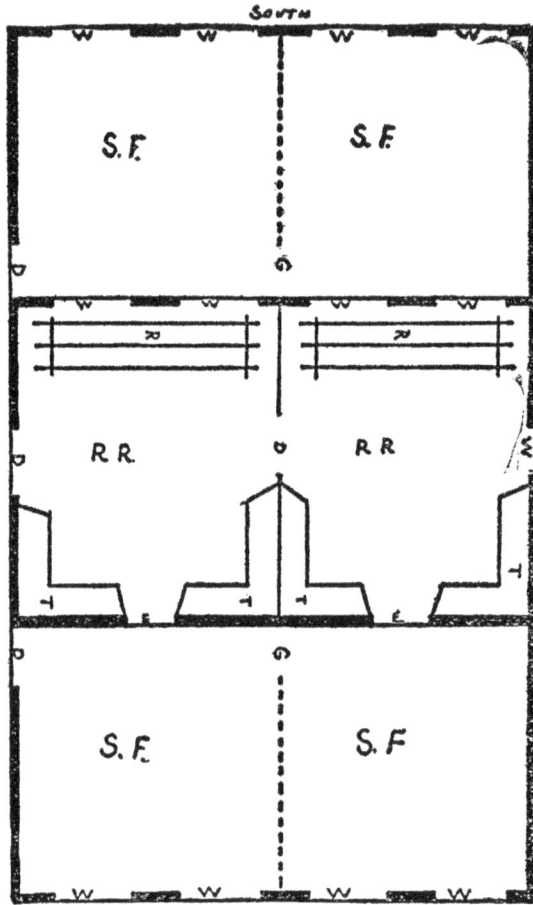

Ground plan poultry house No. 5.

D, Doors.	R R, Roosting room
E, Hen doors.	S F, Scratching floor.
G, Gates in partition.	T, Feed troughs.
M D, Manure door.	W, Windows.

24

The space under floor of main building is a part of scratching floor with a lath partition extending through. It is calculated as an all-year poultry house; the windows in north and south are hinged and are intended as outlets. This house is suitable for 50 hens and two males. The object in having it divided is to allow of the flock being separated into two lots in case of a bad spell of weather when hens should be enclosed. It is not necessary to separate them, but I do it, as I think when they are confined to the house for weeks at a time it is better to have the flock divided.

BROOD COOPS.

If satisfactory results are to be had, when chicks are reared by their natural mother, the hen, a good brood coop is almost a necessity. To be sure one can get along with old boxes and barrels if he will be ready at any moment to dash out and secure them against danger on the approach of the sudden wind and rain that is sure to come during the spring and early summer, or to seize lantern and shot gun at the first *squawk* when the festive skunk pays a visit to the range. But the question is, does it pay to be a prey to all this anxiety and perhaps in the end lose a large portion of the young stock? We do not think it does, and hence illustrate, in this issue, a brood coop which can be built by any poultryman and at small expense. No. 1 shows the completed coop as it would appear when occupied by a hen and brood. The dimensions of this coop are as follows: Bottom, 18 inches wide by 24 inches long *on the inside* (the siding should extend down *over* the edge of bottom instead of resting *on* the bottom, as shown in illustration; this will prevent rain from beating in at this point and wetting the bottom of the coop). The sides should be 12 inches high at rear and 21 inches at front, to give proper slope to roof. The roof is made separate from coop and nailed to two cleats about 1x1½ inches, notches are cut in the sides of coop to fit these cleats which have a small hook at each end, staples are driven into the sides of coop, at proper place, and roof can then be hooked on secure against any wind. The object in having roof movable is for convenience in cleaning. It can be quickly and easily removed, coop turned upside down and a few taps on the bottom will clean the coop. The loose roof or "flap" that is shown as a "porch roof" is one of the greatest advantages of this coop, as it may be arranged to suit the requirements of all kinds of weather, and let down entirely and fastened at night, thus securing the hen and her brood against molestation from without. This flap is shown in illustration as being fastened with iron hinges, but a strong piece of leather may be used instead and will allow "flap" to be laid entirely back upon the roof or closed entirely as may be desired; inch cleats should be nailed across this flap also, thus allowing an inch space for ventilation when same is let down over the front of coop.

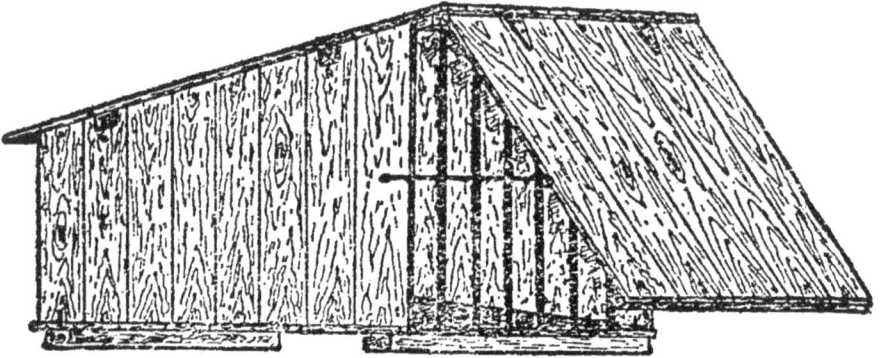

No. 1.

The cleats shown on bottom of coop are about 1½ x 1½ inch, and raise the coop above the ground sufficiently to prevent any dampness. A slatted door, to fit the front of coop may be made and fastened at one side with hinges, and will be found very convenient when letting hen out or in.

This coop may be made of ⁵⁄₈ matched stuff, and will then be light and easy to move about, the roof should be well painted or covered with roofing paper and then painted, and should extend *over the sides* of coop about three inches. The material for such a coop will not cost to exceed 60 cents, while the saving in anxiety, "fussing," and valuable chicks will repay the first cost many times during one season, and if coops are well made and stored away carefully when not in use, they will last for years.

A small yard, two feet wide, four feet long, and as high as the front of the coop, and made of lath, will be found a convenient addition to this brood coop, as the hen may be allowed to be outside of coop and yet not run at large, while at feeding time the chicks will be secure from the older fowls and can get the full benefit of the rations given them.

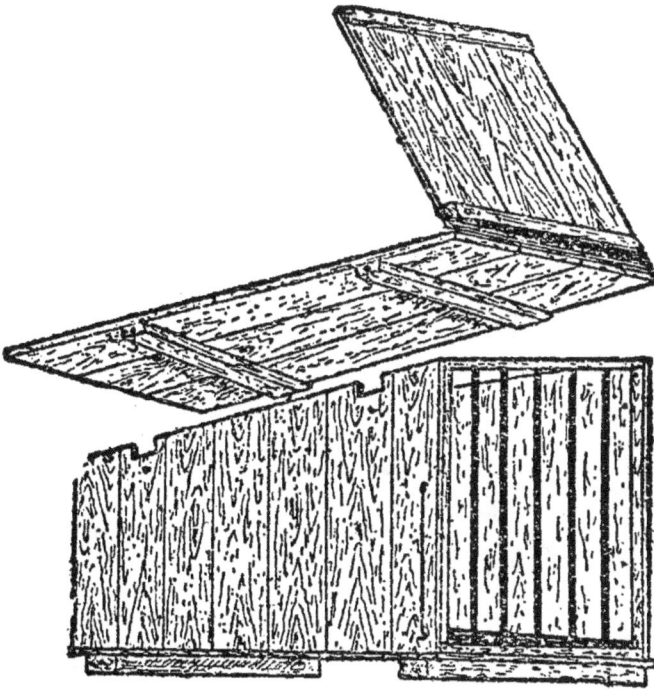

No. 2.

To arrange coop for the reception of the hen and chicks just hatched, place it (if in early spring) in some sunny spot, on a slight elevation if possible, put in dry dirt sufficient to cover the bottom of coop to the depth of about half an inch, cover with fine, dry straw or chaff, and coop is ready for tenants. The dry dirt in bottom will absorb all moisture and prevent filth from adhering to the coop. On cool, sunny days turn the "flap roof" back and allow the sun to shine in. When it is wet and stormy, the same may be let down as shown at No. 1.

In brood coop Fig. 3 we have the old fashioned one that has been in use for centuries and to this day has not become extinct. This is quickly made, and from almost any kind of material that may happen to lie around the place. The end pieces are nailed together at the top cone-shaped or much like an inverted letter V. Then take old barrel staves, or any old boards that may happen to be lying around, nail them on for a cover, leaving one loose at front side for a door, as shown in the illustration. This door may be fastened with leather straps which will serve as hinges. Best make a bottom out of boards to set the coop on after it is completed. This will prevent the chicks from being drowned in case of heavy rain. This bottom should have cleats one inch thick under it so as to let all water flow through under the coop. In the early spring when the weather is cool, all coops should be placed in the sun where it is warm, but after the weather grows warmer they must be removed to the shade, as chicks cannot stand too much heat. Too much heat is productive of bowel disease in chicks, the same as becoming chilled—both produce the same results. Always provide abundance of shade in very warm weather and abundance of sunshine in quite cool weather.

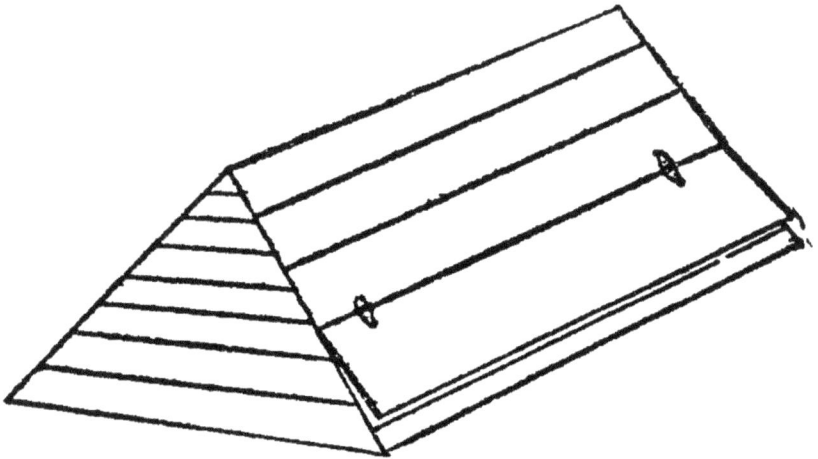

Fig. 3.

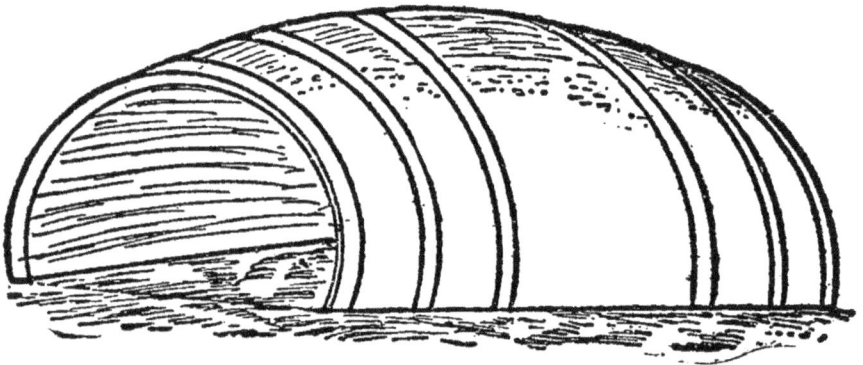

Fig. 4.

In Fig. 4 we have an illustration of a barrel sawed in two lengthwise, the staves first being nailed to the hoops so that it cannot fall together while being sawed apart. This coop is roomy and cheap, and any one can get a sugar or a flour barrel to make a brood coop. A false bottom may be provided for the barrel coop, the same as in Fig. 3, which will serve as a protection against the chicks getting drowned in case of a heavy rain. In front of the barrel, boards three inches wide may be driven in the ground, far enough apart to permit the chicks to run in and out. The middle board may be so arranged that it may be removed in order to let the mother hen out when desired.

In Fig. 5 we have a very nicely arranged coop which is a little more expensive, yet it can be constructed from very cheap material and the cost will be very small compared to its convenience. In place of the slide door a loose slat may be removed to let the mother hen out with her brood. There is a false bottom to this coop which extends 8 inches in front of the coop, on which to place the drinking fountain for the chicks; and the mother hen can also drink by reaching through between the slats, which are set 2½ inches apart.

Fig. 5.

You will notice that the sides of this coop extend down over the edge of the bottom and rest on the cleats that are nailed fast crosswise. This device prevents any water from running into the coop; hence the bottom is kept perfectly dry at all times. The roof of this coop should extend 2 inches over the sides, though as you see it in the illustration, the sides are nailed to the side of the roof (an error of the artist). You notice the illustration indicates the side as being nailed to the bottom; this also is wrong. The sides drop down loose on both sides, facilitating the cleaning of the coop.

All coops should be cleaned at least once a week and the floors saturated with kerosene oil. This, if properly done, will keep them from being overrun with lice during the warm weather.

Fig. 6.

Fig. 6 is only a store box closed all around except one side, and on this side are 3-inch slats nailed 2½ inches apart. This may be set in front of any of the brood coops to allow the chicks to exercise during the first week after coming from the shell. I have found this method very practical. They should not be allowed to run out in the morning until the dew is entirely off the grass, and this extra compartment serves well for the purpose of giving more room until time for them to be liberated.

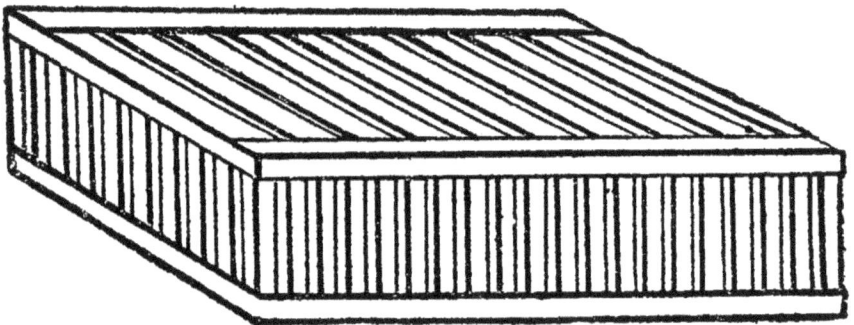

Fig. 7.

In Fig. 7 we have a feeding coop which is constructed entirely from 3-inch slats. This device is intended to be used for feeding the young chicks when they are from 4 weeks to 2 months old. The coop may be set where it is most convenient for the chicks. All food may be thrown in this coop, the slats being only far enough apart to permit the young chicks to enter, thus preventing the old hens or larger fowls from eating the food from the younger or smaller ones.

This feeding coop may be made either large or small to suit the flock of chicks you are caring for. Or you may have a number of them, as it is not best to make them so large as to be unhandy to move from one place to another.

CHAPTER IV.

SELECTION OF BREEDS AND BREEDING.

A mistake is oftentimes made in selecting fowls of a breed that is not suited for the purposes for which they are to be kept. If egg production is the all-important point, it is a most serious mistake to select a breed of fowls that is not noted for this product. If, on the other hand, meat is the chief object, an expensive mistake will be made if any but heavy-bodied fowls are chosen. Then, too, if fowls are kept for both meat and egg production, some breed of the middle class should be chosen. These, while they do not attain the great size of the Asiatics, are sufficiently large to be reared profitably to supply the table with meat, and at the same time have the tendency for egg production developed sufficiently to produce a goodly number of eggs during the year. The Wyandottes and Plymouth Rocks are good illustrations of this class of fowls. While individuals of these breeds have made excellent records in egg production, the records of large numbers do not compare favorably with the egg production of the Mediterranean fowls. All of the so-called Mediterranean fowls have a great tendency toward egg production, and require only the proper food and care to produce eggs in abundance.

A serious mistake is also made in selecting fowls for breeding purposes and in selecting eggs for hatching. On many farms the custom is to select eggs for hatching during the spring months, when nearly all of the fowls are laying. No matter how poor a layer a hen may be, the chances are that most of the eggs will be produced during the spring and early summer months. A hen that has laid many eggs during the winter months is quite likely to produce fewer eggs during the spring and early summer months than one that commenced to lay on the approach of warm weather. Springtime is nature's season for egg production. All fowls that produce any considerable number of eggs during the year are likely to be laying at this time. It is therefore plain that whenever eggs are selected in the springtime from a flock of mixed hens, composed of some good layers and some poor ones, a larger per cent of eggs will be obtained from the poor layers than at almost any other season of the year. A serious mistake is therefore made in breeding largely from the unprofitable fowls.

Whenever it is possible, fowls that are known for the great number of eggs they have produced during the year should be selected for the breeding pen. While it will be almost impossible, and certainly impracticable, in the majority of cases to keep individual records of egg production, yet a selection may be made that will enable the breeder to improve his flock greatly.

The two things necessary to produce large quantities of eggs with the Mediterranean fowls are (1) Proper food and care, and (2) a strong constitution, which will enable the fowls to digest and assimilate a large amount of food; in other words, fowls so strong physically that they will stand forcing for egg production. In this relation, we may look at the fowl as a machine. If that machine is so strong that it can be run at its full capacity all the time, much greater profit will be derived than if it can be run at its full capacity only a part of the time.

There is, perhaps, no time in the history of the fowl that will indicate its vigor so well as the molting period. Fowls that molt in a very short time and hardly stop laying during this period, as a rule, have strong, vigorous constitutions, and if properly fed give a large yearly record. On the other hand, those that are a long time molting have not the vigor and strength to digest and assimilate food enough to produce the requisite number of eggs. If it is necessary to select fowls at sometime during the year other than the molting period, some indication of their egg-producing power is shown in their general conformation. It is a rule that fowls bred for egg production are larger bodied than those bred for fancy points. Whenever vigor and constitution form an important part in the selection of fowls for breeding, the size of the fowls is invariably increased.

IMPROVEMENTS OF BREEDS.

The improvement of breeds is a subject that has received most careful study from scientific men. It should receive the attention of all poultry raisers. Either one's flock will grow poorer or it will remain indefinitely as it is, or it will increase in productiveness and so in profit. Now there is not likely to be any increase unless it is brought about by most careful attention, especially to the subject of breeding. Good care except in this particular and the general operation of the law of heredity, will keep the flock at about the same value for a long period; while neglect in the general care of a flock will cause it to decrease in value.

Heredity produces uniformity. Improvement is to be sought only in variation. Some animals will be found to have a capacity for variation from generation to generation, while others will change but little. The breeder who wishes improvement always looks for variation. If any line of descent shows marked variation, he at once begins to experiment with that in the hope of improvement. Of course the variation may not be for the better. That is a thing one must take chances upon. But at any rate the only chance of improvement is in variation, and in selecting and breeding fowls that show a tendency to variation. As a rule, males vary more than females, and the young more nearly resemble the female than the male.

IN-AND-IN-BREEDING.

Another method of producing better fowls is what is known as in-and-in-breeding. It consists in mating closely related fowls of superior value. Thus if one has one or two exceptionally good layers, their chicks will be mated with the old fowls, or with others from the same stock, thus increasing in a progressive ratio the blood known to be good.

There is some difference of opinion as to in-and-in-breeding, some holding that too much mating of near relatives produces deterioration. There is no doubt some truth in this. At the same time, there is probably less objection to the practice in the case of fowls than of any other animals. In-and-in-breeding has played an important part in the production of breeds, and there can be no doubt that carried to a moderate degree it is eminently useful. Of course there will be examples of deterioration. Not all experiments of this kind will result successfully, but these exceptions will be found in all methods of improvement by breeding.

It is sometimes supposed that by mating two fowls of superior qualities, the good qualities of both will be obtained in the offspring. This is by no means always the case. Often the bad qualities of both will be perpetuated, while the good qualities will be lost. However, good qualities differing in kind are sometimes doubled up in this way.

CROSS-BREEDING.

Cross-breeding, or the mating of two well known and distinct breeds with each other, except for special purposes, is not as popular now as it once was.

Very often in cross-breeding the progeny will revert to the type of some early ancestor. Darwin speaks of mating a black Spanish and a White Silky and getting a fowl much resembling the wild Jungle Fowl of India, supposed to be the source of all our domestic birds. Reversion is indeed very often likely, to occur when a cross is made between very distinct and well established breeds. The blood of the two breeds does not blend well. But cross-breeding is responsible for new breeds in many cases, as for instance the Plymouth Rocks and the Wyandottes.

DO THOROUGH-BREDS PAY?

The question is often asked if it pays to raise thorough-breds, or pure-bred stock. The answer is that everything depends on the care the animals are to receive. There is no doubt that thorough-bred stock does pay if it is properly cared for; but if it is not well cared for it will soon relapse into the very ordinary kinds, or something even less valuable. So the attempt to raise pure-bred stock will result only in a loss. The common stock of mixed-breeding is usually more hardy, and on the whole is better adapted to care for itself where the fowls are turned out to scratch for themselves and no special attention is paid to them.

Food has always, and always will, play an important part in the improvement and perfection of breeds. The raiser of common fowls who says he does not believe in pampering his animals makes a sad mistake, and at least for him thorough-breds have little or no value.

And yet the scientific man can undoubtedly get a great deal more out of improved breeds if he will take the time and trouble to care for them properly, and has the brain to do so. Well bred stock should always have an abundance of the proper food, warm houses, and clean pens. There is no doubt that they pay when well cared for.

CHAPTER V.

THE DIFFERENT BREEDS.

For convenience we may divide fowls into four classes, namely:
Egg breeds.
Meat breeds.
General-purpose breeds.
Fancy breeds.

Of course this division is not hard and fast, for one breed may be considered by one raiser as a meat breed, and by another as a general-purpose breed because it has good egg-laying qualities. Every man must judge for himself, as the result of experiment.

In a general way the egg breeds include the small and nervous fowls that have a strong tendency to produce eggs all the year round. They are as a rule of little value as sitters, and while young at least they are too active to fatten easily. Leghorns Spanish, Minorcas, and Hamburgs are good illustrations of this type at its best.

The meat breeds are usually the largest kinds of fowls, being much larger than the egg breeds, and somewhat larger than the general-purpose fowls. They make the most persistent sitters, being large, slow, and gentle, and not easily frightened. To this class belong especially the Brahmas, Cochins, and Lanshans.

The general-purpose fowls are fairly good for both meat and eggs. Under good conditions they will lay fairly well, and they are of fair size and afford a good quality of meat when it is properly prepared. Plymouth Rocks and Wyandottes are the best illustrations of this class.

The fancy breeds are raised more for looks than for utility, and are typified by the Polish and Bantams.

LIGHT BRAHMAS.

The Light Brahmas are one of the oldest breeds on the poultry list. They have been bred in the Old World centuries ago. All through the annals of the history of poultrydom they have figured prominently as leaders in their race, and withstood trying ordeals as no other breed ever has. New breeds may come and go, but the good old Light Brahma keeps pace with the times. They have maintained the foremost position among thousands of scrutinizing admirers down through all ages, and they continue to satisfy all who have tried them. Any breed giving universal satisfaction among so many breeders must have qualities of a very high order. Qualities that are undisputed and rarely possessed in any other variety! They are very large, dress well for market, have a nice yellow skin, will fatten nicely when matured, and they command the highest market price. Eggs are quite large and brown in color. Some are lighter in color than others. The best specimens sometimes laying the lightest colored eggs. The females make good sitters and good mothers. They will thrive well in small enclosures. A fence four feet high will be sufficient to keep them.

Light Brahmas.

DARK BRAHMAS.

Dark Brahmas.

The Dark Brahmas are one of the most prominent members of the great Asiatic family. They are very beautiful, especially the female. The plumage of the female is a steel gray with delicate pencilings, except on hackle, where the pencilings are quite prominent, making them a very desirable fowl for the city and town, as the dust and soot will not soil their plumage. The plumage of the cock, although quite different from that of the hen, commands admiration of those who have a taste for the beautiful in nature. No breed is more hardy from the time it picks the shell until ripe in old age than the Dark Brahma. As egg producers they are second to none of the Asiatics. With their great vigor of constitution when young, they feather rapidly and are ready early for the market as spring chickens or broilers. In weight they are about the same as the Light Brahma, cocks weighing at maturity, when in good condition, from 10 to 12 pounds, and when in good condition, 8 to 10 pounds.

BUFF COCHIN.

The Buff Cochin is the oldest of the Cochin family. They are pure and princely Asiatics, coming from the Orient—the starting point of all good, where we find the first habitation of the human race and the beginning of the Christian era. Dr. Spaulding says the Buffs were once the awe-inspiring Shanghai, whose clarion crow shook the western continent some 50 years ago. The Buff Cochins are elegant fowls, being compact, good layers, good sitters, good mothers, and are well adapted to confinement in small enclosures or yards.

Buff Cochins.

They are very large and broad, ranking in size with the Brahmas, are very hardy, both as chicks and fowls, it being seldom that sickness of any kind is found among them. They are well adapted to cold climates, being quite heavily feathered, and not having very large combs and wattles, so they are not apt to get frost-bitten in the severest weather. In color they are a rich buff, which makes them a suitable fowl for country, village or city. They are of very quiet habits, will not fly over a fence 4 feet high, which makes them very desirable, as it insures comfort to your neighbors, and you have the satisfaction of living in peace with those around you. They are good winter layers. Early hatched pullets will begin laying in December and lay all winter.

Size of cocks, when matured (in good flesh) will weigh from 9 to 12 pounds. Hens from 7 to 10 pounds; however, in many instances they have attained much greater weights than those given.

WHITE COCHINS.

White Cochins.

In a general way the White Cochins are the same as all other Cochins excepting in color. The above cut gives a good representation of this grand old breed of fowls. They possess the same good qualities as the Buff Cochins, and in our long experience in breeding we find them exceptionally good layers for a large fowl, and we do not hesitate to recommend them as being better layers than some of the other Cochin families. They are very large and have a stately appearance, are quite prolific and vigorous growers; being very hardy they withstand disease better than many other varieties of fowls. Chicks mature rapidly, which is just what is needed for a valuable market fowl. There is no breed of fowls better adapted for small enclosures than the White Cochins, as they bear confinement exceptionally well; but if allowed their liberty they are very good foragers.

PARTRIDGE COCHINS.

Partridge Cochins.

The Partridge Cochins are the most popular and to us the handsomest of all Cochins. With many it is the favorite breed. In plumage they are rich and elegant, and so dark as not to become soiled when kept in city yards. The hen is a rich brown, with beautiful cross pencilings in black; hackle, golden or yellow, striped with black, having a downy appearance and a satisfied, motherly bearing. Cocks have solid black breast, back red, hackle and saddle orange red, with fine, well-built, symmetrical form and proud, aristocratic carriage. They are large; cocks weigh 9 to 12 pounds, hens 7 to 10 pounds at maturity. They are easy to rear, extremely hardy, breed remarkably true to feather, are very fine in shape, have yellow legs and skin and sell well in market.

41

Excellent layers during the greater part of the year, and their flesh is toothsome, being tender, juicy and presentable in color. They are good sitters and good mothers. They are eminently fitted for either the farmer, cottager, fancier or mechanic, or anyone desiring a large and beautiful fowl, quiet and gentle in disposition, and not inclined to roam.

BLACK COCHINS.

Black Cochins.

Black Cochins are not bred quite so extensively as the other Cochin varieties, which is probably due to their color. Their general make-up is the same as that of other Cochins, and they possess quite rare qualities; are very large, weighing at maturity, when in good flesh, as high as 12 pounds. However, this weight is only attained by some of the finest specimens. Have nice yellow skin, which for market purposes is preferable. They are good layers of nice dark-colored eggs, and they lay better during the winter months than many others. They make good sitters, kind and gentle mothers, giving the best of care to their young. Chicks grow rapidly, are strong and hardy. Will mature in about six months. The Black Cochins are easily confined to small inclosures, unlike in any other breeds; when allowed to roam they make good foragers.

BARRED PLYMOUTH ROCKS.

Barred Plymouth Rocks.

The Barred Plymouth Rocks are so well known the world over that it is almost useless to give them a detailed description. They are found among the breeds of all fanciers generally, and it is with pride that we can point to this variety of fowls, and claim it to be a pure American breed, with qualities for an all-purpose fowl equal to the best. They are vigorous, noted for hardiness of constitution; young chicks grow rapidly and are ready for market at a very early age. When matured they are large, ranking next in size to the Asiatics; they have clean legs, beautiful blue-barred plumage, are first-class layers, good mothers, but not inveterate sitters, and an excellent table fowl. This breed is undoubtedly one of the most profitable breeds in existence, and is acknowledged the farmer's fowl. As to market quality, we need not call your attention, as the name Plymouth Rock is all that is necessary.

BUFF PLYMOUTH ROCK.

Buff Plymouth Rocks.

Still the new breeds continue to come, and as long as they come with merits equal to the Buff Plymouth Rock, we have room for them. This new breed deserves the attention of all who are interested in a fowl with such merits as are found in the other varieties of the Plymouth Rocks. They possess the same general characteristics as do all their ancestors, with the exception of color. This is a new breed, its origin dating back to only a few years ago. While not an old variety, we are glad to note their breeding qualities as being superior to most others. The merits of the Buff Plymouth Rock cannot be disputed, as they are in the strictest sense a first-class all-purpose fowl. They are quite large, have nice, well-rounded bodies, and a bright yellow skin. They have clean legs of medium length. Are well adapted to both the fancier and the market poultryman. In our experience we find them better layers than the Black

Plymouth Rocks. Their eggs are of about the same color, size and quality as other Plymouth Rocks. They are good sitters and make excellent mothers. Their young are vigorous growers and round up at an early age.

WHITE PLYMOUTH ROCKS.

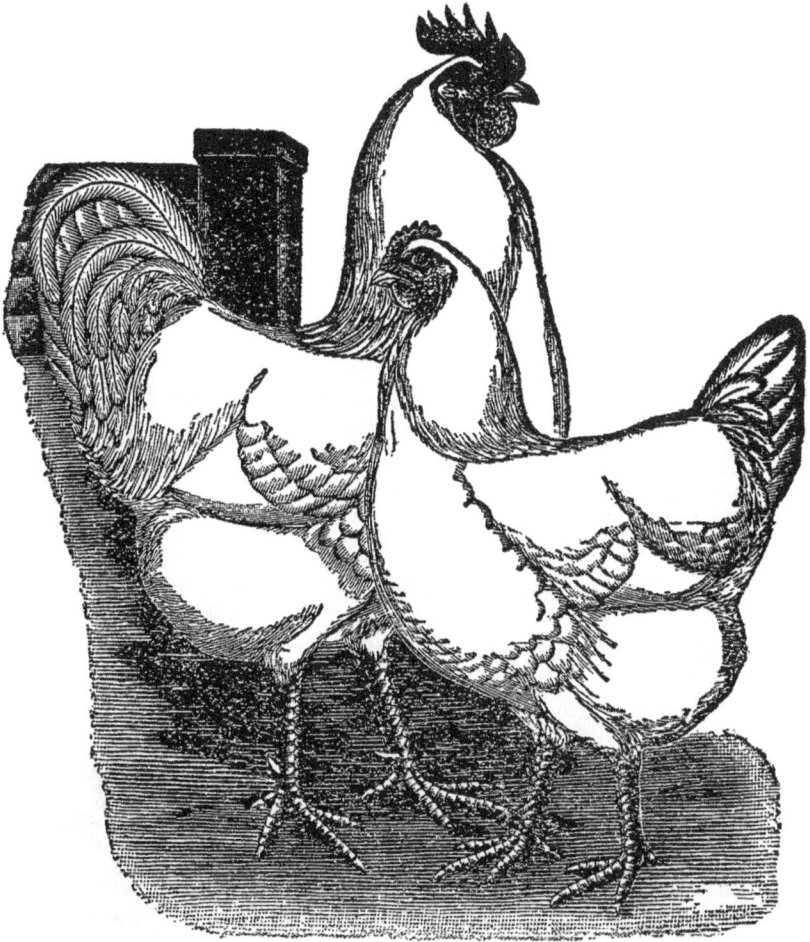

White Plymouth Rocks.

The White Plymouth stands to-day, without any exception, at the head of all general purpose fowls. They are exceedingly hardy and mature very early. Standard weight for cocks, 9½ pounds; hens, 7½ pounds. They dress excellent, having a fine yellow skin and legs, which are admired very much in market poultry. They are of fine build, very stylish, and one of the best laying breeds in existence, laying mostly brown-shelled eggs, although the color of them varies from light to a dark brown. For the market poultryman we can recommend no breed suited better to his wants than the White Plymouth Rocks. Our stock is second to none, scoring from 90 to 95 points.

Our fowls are of fine build, large and heavy, and stately carriage; we would invite you to give them a trial and be convinced for yourself of the rare qualities which the White Plymouth Rocks possess.

PEA COMB BARRED PLYMOUTH ROCKS.

Pea Comb Barred Plymouth Rocks.

The Pea Comb Barred Plymouth Rocks have been bred from the Single Comb variety. They have been bred from sports; thus they retain the excellent qualities of their ancestors. They are quite large, being about the size of the Single Comb Barred Plymouth Rocks; but they have a pea comb, which is much desired in our colder Northern climates on account of its being nearer frost-proof than larger single combs. They possess the best qualities to be found in a general market fowl. Are large, have a nice yellow skin, smooth legs, which should be yellow, and in most cases are yellow. However, some few dark spots on the leg will not disqualify, but are simply imperfections. They have not been bred as extensively as their single comb ancestors, consequently they are not bred as fine in plumage as the single comb variety. They are splendid layers of mostly brown or yellow eggs. Some are lighter in color than others. They make fine mothers, and are good sitters, yet they are not as persistently broody as some of the larger varieties. The chicks thrive well with ordinary care, and will grow to weigh a pound and a pound and a half at an early age. They are a breed that we can recommend to anyone who wishes to have a good valuable fowl for their own use. Our matings in this variety are exceptionally fine this season. The males are large and well barred to the skin. Hens and pullets of this seasons matings are also large and well-marked birds.

PEA COMB WHITE PLYMOUTH ROCKS.

Another variety of white fowls which bids fair to become very popular among the masses are the P. C. White Plymouth Rocks. Like the single comb variety, they possess all the desirable qualities of an all-purpose breed, and in one point they will surpass them. Having the pea comb, they are what may be called frost-proof, which is a very desirable feature in cold climates. They are not bred as extensively as the single comb variety, owing to the fact that they are practically a new breed, and people have not become acquainted with their qualities as they should. They are large, good style, grand, stately carriage, well-shaped body, have a nice yellow skin, which is so desirable in market poultry. Hens make good sitters (but are not persistently broody), good mothers, will care for their young as well as any breed. Their chicks grow very rapidly, feather out early and will round up for market when quite young. Their laying qualities are not to be disputed. They will lay as many eggs as any of the large breeds. Their eggs are the same in color as other Plymouth Rock eggs. This is a noble breed, and is destined to become one of the leading varieties when once their merits are better known.

Pea Comb White Plymouth Rocks.

CORNISH INDIAN GAMES.

This is a breed that is gaining public favor very rapidly. It is a breed for general purposes, having the qualities of a market fowl, *i.e.*, compactness, yellow legs, heavy weight (cocks, 9 to 11 pounds; hens, 7 to 8½ pounds), from which there is but little waste in dressing, and, being of quick-growing habits, they produce a fine broiler in a short time because they have short feathers, the nutriment required to put feathers on other breeds going to flesh, which is more juicy and tender in this breed than in a young turkey. The Indian Game hens are good sitters and mothers, and the chicks are very hardy. In color the fowls are quite pretty; the bright brown shafts and the glossy green lacing make a beautiful contrast. Here we have a fowl for the market poultryman, the farmer, the broiler raiser, the egg producer and the fancier. The general appearance is that of a powerful bird. Body very broad, thick and compact; flesh firm and solid. In the male the plumage of the breast and underbody is a green, glossy black; neck-hackle same color, with brown crimson shafts to feathers; back and saddle a mixture of green, glossy black and brown crimson. Wings chestnut brown, with metallic green, glossy black wing bar. In the hen the ground color is chestnut brown, with beautiful lacings of metallic green, glossy black. The legs and skin in both sexes are a very rich, bright, deep orange yellow, which makes them very desirable for market. Face, wattles and comb are a rich red.

Cornish Indian Games.

RED PYLE GAME.

The Red Pyle Game is in general make-up and quality the same as the Brown Red or Black-Breasted Red Game, only differing in color. They are very courageous and hardy, and to most people not lacking in beauty. Color of cock, hackle and saddle orange, light red or chestnut; breast, shaft and margin of feathers chestnut red, wings white and red, tail white, body white. The hens are mostly white to a creamy white, running darker on breast and wings.

Red Pyle Games.

BLACK SUMATRA GAME.

The Black Sumatra Game is not as well known in the poultry circle as many other varieties of the Game; yet we have bred them right in line for many years, and by a thorough test we find them to be a fowl deserving of credit. They do not resemble the Game family to a very great extent, being more heavily feathered than most Games. In color they are a solid black throughout, shading being a lustrous green, which makes them very rich and handsome in color. They are very good layers of medium-sized eggs, ranging in color from white to a darker shade. They make exceptionally good mothers, and will take care of a brood of chicks as well as one might wish. Chicks are hardy and grow fast. They are a splendid table fowl, as all Games are. Cocks weigh, when in good flesh, from 6 to 8 pounds, and hens, from 5 to 7 pounds.

Black Sumatra Games.

51

BLACK-BREASTED RED GAME.

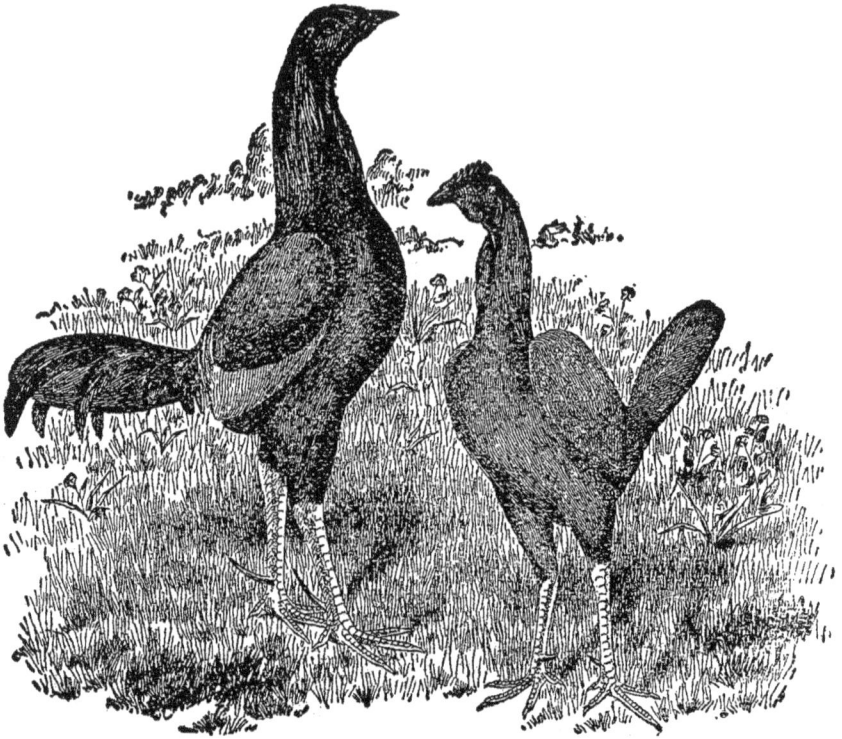

Black-Breasted Red Games.

The Black-Breasted Red Game is the best known, being the oldest breed and probably bred the most extensively of any variety of the Game family. They stand at the head, and they justly deserve their prominence. It is hardly necessary to give a full description of this old breed, as it is so well known. Now we come to a difficult point; as all Games possess good qualities, and in size they are all about the same, except the Indian Game, they being somewhat heavier than other Games. They all possess courage and are very hardy and vigorous, but the B. B. R. Game is the standard fowl in the Game circle. In color they somewhat resemble the Brown Leghorn, but are larger than the Leghorn. The Games are all good mothers and good sitters, but not inveterate hatchers; and as to the laying qualities of the Games, they are all the same. One variety lays about the same as another, but they are not supposed to lay as many eggs as the Leghorn. They range in that line about the same as the Plymouth Rock, Java or Wyandotte. Young chicks feather very rapidly and grow fast.

BROWN RED GAMES.

Brown Red Games.

The Brown Red Games are second to none in this country—fine specimens, with great vitality and hardiness of constitution. To say they are good layers will not do them justice, as they have proved themselves in laying almost equal to any fowl, and surpassed by few. They are very handsome, the color of male, neck, back and saddle lemon, with narrow stripe of black in the middle of the feathers; breast, ground color black, laced some with lemon; wing bow lemon, wing coverts glossy black, primaries and secondaries black and tail glossy black. The female, neck lemon, with a narrow stripe of black in middle feathers; breast, ground color black, evenly laced with lemon; otherwise the plumage is black throughout. They bear confinement remarkably well, are hardy both as chicks and fowls. Chicks grow very fast and mature quite early. The hens make splendid mothers, taking the best care of their young. Cocks weigh from 6 to 8 pounds; hens from 5 to 7 pounds. They are a fine table fowl and dress very nice and presentable.

BLACK LANGSHANS.

The Langshan belongs to the old Asiatic class of fowls. No variety has risen to distinction and prominence more rapidly than has the Langshan. They have gained the entrance to the front ranks in the poultry fraternity. Langshans were imported to this country about twenty-five years ago—origin claimed to have been in the northern part of China. They are very prolific, thrive well in either small inclosures or running at large. The chicks when first hatched are about half white, and quite frequently they will show some white feathers in wings after they are feathered out, but after they molt these chick feathers, black feathers will come in where the white ones were seen. In color the Langshan is a beautiful green, glossy black throughout, which makes it decidedly handsome.

Black Langshans.

Now, as to their laying qualities, we have found them excellent layers, better than the average large fowl. Early hatched pullets will begin laying in early winter and continue throughout the whole winter, and will lay reasonably well all spring and summer. However, the condition and care of fowls must be taken into consideration, not only in the Langshans, but in all breeds. They lay mostly dark colored eggs, yet they do not lay strictly one color. Some are darker in color often than others, which is no indication of impurity of stock. They are very large in size and well built.

BUFF LANGSHANS.

These are a new variety of Langshan, differing from others merely in color. They are an excellent general purpose breed.

Buff Langshans.

BUFF COCHIN BANTAMS.

The Buff Cochin, or Pekin Bantams, in color are a golden buff same as the Buff Cochin, except that the cock is usually dark, almost chestnut. They are very quiet, tame and docile, making the very finest pets. In size they are just a little larger than the Golden Sebright Bantams, being a little broader and heavier set. All who see them admire them for their exquisite, unique appearance and beauty. They are quite good layers of rather small eggs, light in color, are good sitters and good mothers. Chicks are easily raised, are hardy both as fowls and chicks. They bear confinement well, thrive just as well in small inclosures as running at large, and are fine pets to have on a nice green lawn where they command the admiration of all who see them.

Buff Cochin Bantams.

GOLDEN SEBRIGHT BANTAM.

The Golden Sebright Bantam is one of the most attractive breeds of all Bantams. No breed among the beautiful pigmies excels the Sebrights in beauty of penciling, and none are prettier pets. In color they are rich golden yellow, laced with black markings, being very distinct and clear. They are very hardy both as chicks and fowls, breeding very true to color. They are both useful and a true fancy fowl, excellent layers of small eggs, good sitters and good mothers. They are quite small, standard weight for cock 26 ounces, hen 22 ounces. They are stylish, active, and no lover of the beautiful can afford to be without them.

Golden Sebright Bantams.

AMERICAN DOMINIQUES.

American Dominiques.

The Dominique, in color, resembles the Black Plymouth Rock, but is not as large, has a longer tail and a rose comb, while the Plymouth Rock has a single comb. The Dominique is one of the oldest varieties, and a pure American breed. They are very hardy, chicks grow rapidly and mature early. Pullets often begin laying at from five to six months old. The hens make excellent mothers and splendid hatchers; not being so clumsy and heavy, they seldom break an egg while sitting. They are one of our best layers of eggs, in color from a light to a darker brown, medium in size. They are a splendid table fowl, many claiming them superior to all others, having a yellow skin, dress well and command the highest price in market. They are very gay, stylish and fine in appearance, are well adapted for confinement in yards, and if left to roam at will they are good foragers. The color of plumage being the same as the Plymouth Rock, they are adapted to all localities, either city, village or country, as the soot, smoke and dirt will not mar their appearance. For general utility they have few, if any, superiors. In weight they are large enough for most people, cocks weighing from 7 to 8½ pounds (when in good flesh); hens from 5 to 7½ pounds, making a nice size fowl for the average poultryman or farmer.

BLACK JAVAS.

Black Javas.

The Java is an old variety of fowls. It is claimed that the Black Javas were bred from the Plymouth Rock. Now, if the Black Java have their origin from the Plymouth Rock, they must of necessity occupy a front rank as an American breed. They will withstand the severest weather in our northern climates. Also are prolific, and thrive just as well in our warmer southern climes.

We have bred the Black Java for many years and find them to be very hardy and vigorous. Chicks thrive well from the time they come out of the shell till old age, maturing very early. We have noticed they are not as apt to take disease as some other varieties, being as nearly proof against disease as any fowl we have ever seen. They are, we think, about the best layers for a large fowl, laying a dark-colored egg, but not strictly one color; some are a little lighter in color than others. The Java hens make excellent mothers, taking the best care of their young. They are good sitters, but are not as persistently broody as are some of the large varieties of fowls.

In color the Black Java is what the name implies, black throughout. The cocks are of a lustrous greenish black about the same as the Langshan. Cocks will weigh from 8 to 10 pounds; hens from 7 to 9 pounds (in good flesh). They possess rare market qualities, having the best possible make-up for a valuable market fowl.

MOTTLED JAVA.

Mottled Javas.

The Mottled Java is a pure American breed of fowls. They possess the same good qualities as the Black Java. They are very hardy and bear confinement in small inclosures as well as any variety. Yet if left to roam they make good foragers. The chicks mature very rapidly and begin laying at from 5 to 6 months old. The hens make good sitters and also good mothers, being very quiet and gentle. They are good layers, as good as any large variety, laying eggs varying from a dark brown to a light color. In size they rank with the Plymouth Rock or Black Java. Cocks weigh when in good flesh from 8 to 10 pounds, hens from 6 to 8 pounds. In color they are black and white intermixed as the name implies (Mottled Javas.)

SILVER LACED WYANDOTTES.

Silver laced Wyandottes.

The great popularity attained by the S. L. Wyandottes in so short a time is without parallel; and no other breed is more highly esteemed in America to-day. They have attained public favor entirely on their own merits, and they are not the coming fowl, but the fowl that has come, and that to stay. For table qualities they cannot be excelled. They are hardy, easily raised, mature early, and for broilers just fill the bill. As egg-producers they are excelled only by the "non -setting" breeds, laying nice rich-colored eggs. They have beautiful plumage, bright yellow legs and skin, low rose combs making them specially adapted to our northern climate. They are of good size, with plump bodies, cocks weighing 8½ to 10 pounds and hens 6 to 8 pounds. The hens are good sitters and careful mothers, but not persistently broody. In fact, this excellent breed combines all the good qualities for a "general purpose" fowl.

GOLDEN WYANDOTTES.

Golden Wyandottes.

In the general utility of the Wyandottes, the Golden, the Silver, the Black and the White are all the same. The only difference being in color, it is a matter of individual taste as to which breed is the best in the Wyandotte family.

They are all splendid layers, and that the whole year around, if properly cared for. They lay eggs in color about the same as the Plymouth Rock. They make the best of mothers, are good sitters, but not persistently broody. Chicks are hardy and grow fast, maturing very early. Cocks when in good flesh weigh from 7 to 9 pounds, hens from 6 to 8 pounds. The plumage of the Golden Wyandotte is the same as that of the Silver Laced, except in color, the center of feathers being a deep rich golden bay, laced with black, giving them one of the richest colors we have in the feathered race.

WHITE WYANDOTTE.

White Wyandottes.

The White Wyandotte needs no detailed description, as the description given the other varieties of Wyandottes will apply equally well to the white variety, except as to color. In color they are white, as the name indicates. They have bright red wattles and ear lobes, comb is rocker shaped, as shown in the above illustration.

They have clean bright yellow legs, and being white they present a beautiful appearance on a nice green lawn. While they are attractive, they possess all the good qualities that combine to make a good all-purpose fowl. Cocks weigh about the same as the other Wyandottes, when in good flesh, from 7 to 9 pounds, and hens from 6 to 8 pounds.

BLACK WYANDOTTE.

Black Wyandottes.

The Black Wyandotte is one of our most valuable breeds, and all because of their good qualities as a general purpose fowl. Their plumage is a beautiful, glossy black, which makes them very attractive. They combine all the good qualities of the Wyandotte family and are more easily bred true to color than their laced cousins, making this variety a very desirable breed to both farmer and fancier.

If you are in search of a good all-around fowl you will find that the Black Wyandotte will come as near to the requirements as any of the older varieties. They are excellent layers of fine, dark-colored eggs; having a nice yellow skin, they make a very desirable fowl for market, and they are just the right build and weight. Hens often reach the weight of 8 pounds, and cocks 10 pounds. The chicks are very hardy and grow fast, making a good chicken for the early market.

BUFF WYANDOTTES.

Buff Wyandottes.

Buff Wyandottes are bred from the other Wyandotte families, and are pure Wyandotte blood even from their very beginning of existence. They are not a new breed of fowls by any means. They have been successfully bred for more than fifteen years, and we have always found them to be exceptionally good layers. When we say good layers we are not doing justice to the breed. They are marvelous layers! They rival any breed on the poultry list, and their eggs are nice size, and in color are from light to a medium dark brown. Are very hardy, and as chicks they thrive and grow very fast under only ordinary care. They make splendid broilers as they feather out very young; and on account of their being a medium weight variety they will fatten and plump up at an earlier age than most other varieties. There is no breed better adapted for broiler raising and for market than the Buff Wyandotte. In size they rank with the other Wyandottes, and they bear confinement as well as any breed on the poultry list. In color they are a nice buff, which is very well suited for an all-purpose fowl, and much admired by all lovers of the feathered tribes.

RED CAPS.

Red Caps.

The Red Caps are an old breed, having been known in England so long that it would be difficult to trace their origin. However, they are an English breed of fowls and are generally known throughout this country. They have a large red comb, from which they derive their name (Red Caps). In plumage the hen is a nut brown, each feather ending with a blue-black spangle. The cocks are black on the breast, but on the back, wing coverts and saddle they are a mahogany red. They are very pretty, bear confinement to small yards quite well, and are considered non-sitters, but we have seen an occasional one gets a little broody, though this is seldom the case. In egg production they are about the same as the Leghorns. They are splendid layers of medium-sized eggs, ranging in color from white to a darker shade. In weight they are larger than the Leghorn, cocks weighing from 6 to 7½ pounds and hens from 5 to 6½ pounds. They breed very true to color, but will throw off single comb specimens occasionally, which is quite common among all rose comb varieties.

HOUDANS.

Houdans.

Houdans are a French fowl. It is not known just how long they have been in existence, nor is it known who the originator was. However, they are a very old breed and are deserving of considerable credit. The Houdan hen is a good layer, her eggs are good size and in color are white. She belongs to the non-sitting varieties. We have bred them for years, and I remember but a couple of instances that any of them wanted to hatch, which is common among all non-sitting varieties. After they are several years old some of them will get a little broody, but are easily broken. The Houdans have five toes, a crest and beard, are shaped much like a Dorking, hence are considered valuable as a fine table fowl, meat tender, juicy and fine flavored. Color of legs of the young is pink, of the old a light gray with sometimes pink on the side. The chicks are decidedly handsome and grow very fast, often weighing four or five pounds at the age of four months, but of course the growth depends largely on the food and care they receive. They are remarkably hardy and thrive well under ordinary care. Standard weight for cock is 7 pounds, hens 6 pounds. They are not a high flyer, will not roam far away, yet are very tame and docile. You can make fine pets of them and will find them well adapted for small runs.

BLUE ANDALUSIAN.

Blue Andalusians.

The Blue Andalusian is an English breed of fowls, dating back fifty years, when they had been bred extensively in Andalusia, Spain, whence they derive their name. They belong to the non-sitting class, and are never known to get in the least broody. As layers they have no superiors, laying the whole year around, both summer and winter. In color the eggs are white and of medium size. They are very hardy, chicks growing very fast, pullets often laying at the age of 4 months; they have red wattles, face and comb, and white ear lobes, are very fine in appearance, gay, stylish, commanding the admiration of all who see them. They are about the same size as the Minorca. However, they are not without a fault, they do not breed as true to color as most other varieties do. They throw off a larger per cent of off-colored chicks than any other variety. However, their other good qualities make up for this weak point.

WHITE MINORCAS.

White Minorcas.

The White Minorcas are supposed originally to have come from Spain. It is not known just how they originated, but as black fowls throw white chicks sometimes it is probable that they originated in that way. As to vitality and productiveness they stand on a level with the Blacks. They are of the same build, have coral red faces, white ear lobes, fully as large in size, and equally as good layers. The White Minorcas (as the name implies) are pure white in plumage, very gay and attractive, having the true Minorca shape and good qualities in general. Standard weight is the same for them as for the Blacks, cocks 8 pounds, hens 6½ pounds.

Black Minorcas.

The Black Minorca is a well established breed of English fowls, belonging to the Spanish varieties, and, wherever bred, are considered a valuable breed; are hardy both as fowls and chicks, easily raised, mature early, and pullets commence laying when very young. They are non-sitters, small eaters, splendid foragers, and without doubt very profitable. Their adaptability to all soils and places, whether in confinement or allowed unlimited range, makes them very popular, and suitable to the city fancier as well as the farmer. Their plumage is a pure black, with a green or metallic luster. Their legs are nice and smooth and medium in length. The chief and striking ornament of the cock is his comb, which is very large, single, straight as an arrow and evenly serrated; has a large flowing tail, carried somewhat high. The comb of the hen lies over on one side of the face, in a peculiar double fold, similar to those of the Leghorn, but much larger. Wattles are in proportion to the combs. The face is red, but the lobes are of a pure white and show up very distinctly. They are very stylish, having a stately, upright carriage, close, compact body, medium low, and are of a stouter and squarer build than the Spanish. Standard weight for cock 8 pounds, hens 6½ pounds.

BUFF LEGHORN.

Buff Leghorns.

The Buff Leghorn belongs to the Mediterranean class, as do all other varieties of Leghorns. They are a new breed, the first ones being imported from England four or five years ago; consequently they have not been bred long enough to get the true buff color in all specimens, but are being perfected each year, so that to-day they breed remarkably well. The principal difficulty is to get a solid buff tail and an even buff color throughout on a male. I have seen but few such specimens. But they are one of the last varieties that we should discard. They stand at the head of the Leghorn family for general utility and are strictly non-sitters. Being but a new variety, they have won for themselves in a very short period the admiration and praise of all leading fanciers. They are promising to outrival their cousins (White and Brown Leghorns) in popular favor, the general make-up of the Buff Leghorns being the same as the other Leghorn varieties. They are very stylish, and have the beautiful buff color, which presents a handsome appearance, making them a favorite variety of fowls with most people. They are great layers, equal to any, if not superior to most other varieties, and in size they rank with other Leghorns; they are small eaters and bear confinement well, and are good foragers when allowed to roam.

SINGLE COMB BROWN LEGHORNS.

The Single Comb Brown Leghorn, of which we present you herewith a good illustration, is so well known that we consider it almost unnecessary to give it a detailed description. Yet peradventure there are some who are not familiar with their habits, we will summarize their merits. First, we would wish to impress upon the mind of the reader that so far as general usefulness, one variety of Leghorns is not in any way superior to another, as they all possess the same general good qualities and usefulness.

Leghorns are considered the best layers, but so far as our experience goes we cannot say they are the best, but they are as good as any; there is no better. The Brown Leghorns have red wattle and comb, white ear lobes, are brown in color, except that the cock is black on breast, deep bay red on hackle. Each feather should have a black stripe in center, back and saddle a deep bay red, tail black. They have a yellow skin and yellow legs. They are very active and spry, and very hardy both as chicks and fowls. Chicks feather out very fast and grow rapidly, maturing at an early age. Their meat is very tender, sweet, juicy and fine grained. Cocks weigh from about 5 to 6½ to pounds, hens from 3½ to 5 pounds.

Single Comb Brown Leghorns.

SINGLE COMB WHITE LEGHORNS.

The Single Comb White Leghorn is in general make-up the same as the Brown variety except in color; they are white. Hence the description for the Brown will answer equally as well for the White. They belong to the Mediterranean class, hence are considered non-sitters; they are excellent layers of eggs, white in color and medium in size. Some Leghorn breeders claim their eggs to be of special fine flavor. However, this depends entirely on what food is given them; if fowls are fed with cabbage, onions or turnips their eggs will have an unpleasant taste. For fine flavored eggs we must feed good, sound, wholesome food, such as corn, wheat, oats, barley, etc.

Single Comb White Leghorns.

The eggs of the White Leghorn usually hatch well, as they are very vigorous and prolific. Chicks are quite hardy and mature at a very early age; pullets often begin laying at the early age of 4 months. They bear confinement well, but are, if left at will, the very best foragers. In size they are the same as the Brown; cocks weigh from 4½ to 6 pounds, hens from 3½ to 5 pounds.

ROSE COMB WHITE LEGHORNS.

Rose Comb White Leghorns.

Rose Comb White Leghorns are identical with the Single Comb variety, except that the comb resembles the comb of the Hamburg. They are much admired by poultry fanciers generally. Their freedom from frozen combs makes them more desirable for our northern climates than the Single Comb variety. They are very stylish and attractive, and belong to the non-sitting class of fowls are splendid layers, as all other varieties of Leghorns. Eggs are white in color and medium in size. They are very hardy, both as chicks and fowls. Chicks grow very fast. Pullets frequently begin laying at 4 months old. They are a fine fowl for the table, as far as they go, but being rather small we could not recommend them as a valuable market fowl, but so far as turning the dollars and cents your way in the egg production they stand in the front rank.

ROSE COMB BROWN LEGHORNS.

Rose Comb Brown Leghorns.

The Rose Comb Brown Leghorns are very popular, being better adapted to our colder climates than the Single Comb varieties. They combine both usefulness and beauty. As egg producers they are simply the same as all other varieties of Leghorns, being first-class in that respect. We have often noticed the statement made about some breeder's favorite variety, as everlasting layers, egg machines, etc. If there are any fowls to which we can justly apply that term it is to the Leghorn family.

They have good style, beautiful plumage, and are, generally speaking, very handsome and attractive. The above illustration gives a good idea of their general appearance. They are hardy, chicks easily raised and mature quite early. Pullets begin laying quite young if well cared for. In size they rank the same as the Single Comb variety. Cocks weigh about 4½ to 6 pounds, hens 3½ to 5 pounds.

WHITE CRESTED BLACK POLISH.

White Crested Black Polish.

The White Crested Black Polish are a very old variety, belonging to the non-sitting class of fowls. They are more bred for fancy than anything else. They are very pretty with their large white crests, yet the crests in many in stances are not entirely white as the name indicates. Nearly all specimens have some black in the forepart of crest, which is only a common thing. However, there should not be much black for a really fine specimen. Aside from their crests they are black throughout. They are splendid layers of eggs, medium in size. They bear confinement well. They are not good foragers. Owing to their large crests, they cannot see well enough, are easily caught by hawks, etc., but for a fancy fowl in town, city or village or country, to keep close around the place or yard, they are, we think, very pretty and ornamental. In size they are about the same as the Leghorn. They breed remarkably well. Eggs usually are fertile and hatch out strong chicks.

GOLDEN POLISH.

Golden Polish.

The Golden Polish are the same as the White Crested Black, they are a very old breed. They are very pretty, the color of the hen being a golden bay, each feather ending with a rich black spangle, or it is laced with black. The cock, on the breast, is a golden bay, each feather ending with a black spangle, or should be laced with black; so you see that they may either be laced with black or be spangled, either being right and in accordance with the standard. They are splendid layers of eggs, medium in size and usually white in color, yet we find many with a yellowish tint. They bear confinement to small inclosures remarkably well, and like other varieties of Polish are not very good foragers, owing to the crest. The Polish generally are not high flyers, hence can easily be kept with an ordinary high fence. The chicks grow rapidly and are remarkably hardy. They mature early and pullets begin laying when quite young. They are very pretty and for fancy and the egg basket we know of no breed that is better than the Polish varieties. In size the Goldens are the same as the White Crested Black.

SILVER SPANGLED HAMBURGS.

Silver Spangled Hamburgs.

The Silver Spangled Hamburgs are one of the most beautiful varieties that can be found on the poultry list. No one can pass a flock of them without a glance of admiration. For beauty they are unsurpassed, and too much could not be said of this beautiful breed. As egg producers they stand in the front rank, laying the year around. In color eggs are white and medium in size. Chicks grow quite fast and mature very early. Pullets often begin laying at 4 months old. They are very small feeders, and bear confinement in small inclosures remarkably well.

For laying qualities and beauty they stand on their own merits and cannot be overestimated. Hens will weigh from 3 to 4½ pounds, cocks from 4 to 5½ pounds.

GOLDEN SPANGLED HAMBURGS.

Golden Spangled Hamburgs.

The description of the Silver Spangled answers equally well for the Golden, with the exception of the plumage. Where the Silvers are white on the body the Goldens are rich, bright, glossy red. The tail is solid black.

The Golden Hamburg is very pretty, one of the handsomest varieties of fowls. They are great layers. Eggs are white in color and medium in size. They are equal to the Leghorns in their laying capacity, are non-sitters, and bear confinement in yards or small inclosures as well as any of the smaller varieties of fowls. The eggs usually hatch well. Chicks if properly cared for grow very fast and mature quite early. Pullets frequently begin laying at the age of 4 months. But they are not considered a valuable market fowl, none of the Hamburgs are. Their merit is in their beauty and egg-producing qualities, which head the list. They are considered an English fowl, but their origin has been at Hamburg, Germany. Their weight is from 3 to 4½ pounds for hens and cocks from 4½ to 5½ pounds.

We have bred them right along in line for a number of years, and consider them profitable as egg producers. Our stock is as good as can be procured anywhere.

WHITE FACE BLACK SPANISH.

White Face Black Spanish.

The White Faced Black Spanish belong to the Mediterranean class, hence are classed with the non-sitting breeds. They have bright red wattles and combs, and face being pure white they present a striking appearance that is characteristic of them alone. Their plumage is a nice glossy black throughout; they are splendid layers of large eggs, mostly white in color, which usually hatch well, as the Spanish possess great vitality. Chicks grow very fast, maturing with good care at an early age. Pullets have been known to begin laying very early and continue laying all winter. They are quite hardy and bear confinement as well as most breeds do. If left to roam they are good foragers. Their meat is fine grained, tender, sweet and juicy. In weight they are about the same as the Black Minorca, viz., cocks weigh from 6 to 8 pounds, hens from 5 to 7 pounds. They breed remarkably true to color, and are a splendid fowl for egg production and trade in general.

CHAPTER VI.

FEEDING.

In feeding for egg production, a valuable lesson may be learned from nature. It will be observed that our domestic fowls that receive the least care and attention, or, in other words, whose conditions approach more nearly the natural conditions, lay most of their eggs in the springtime. It is our duty, then, as feeders to note the conditions surrounding these fowls at that time. The weather is warm, they have an abundance of green food; more or less grain, many insects, and plenty of exercise and fresh air. Then, if we are to feed for egg production, we will endeavor to make it springtime all the year round; not only to provide a warm place for our fowls and give them a proper proportion of green feed, grain and meat, but also to provide pure air and plenty of exercise.

Farmers who keep only a small flock of hens, chiefly to provide eggs for the family, frequently make a mistake in feeding too much corn. It has been clearly proved by experiment that corn should not form a large proportion of the grain ration for laying hens; it is too fattening, especially for hens kept in close confinement. Until the past few years corn has been considered the universal poultry food in America. This, no doubt, has been largely brought about by its cheapness and wide distribution. The recent low prices of wheat have led farmers to feed more of this grain than formerly, and with a frequent improvement in the poultry ration.

When comfortable quarters are provided for fowls the nutritive ratio of the food should be about that 1:4; is, about one part of protein or muscle-producing compounds to four parts of carbohydrates, or heat and fat-producing compounds. Wheat is to be preferred to corn. Oats makes an excellent food, and perhaps comes nearer the ideal than almost any other grain, particularly if the hull can be removed. However, American egg buyers seem to desire dark rather than light yellow yolks, and wheat and oats make light yolks, while corn makes dark yolks. Hence some corn should be fed with the other grain if dark yolks are desired.

Buckwheat, like wheat, has too wide a nutritive ratio if fed alone, and produces a white flesh and light colored yolk if fed in large quantities. In forcing fowls for egg production, as in forcing animals for large yields of milk, it is found best to make up a ration of many kinds of grain. This in variably gives better results than one or two kinds of grain, although the nutritive ratio of the ration may be about the same. It has been found by experiment that the fowls not only relish their ration more when composed of many kinds of grain, but that a somewhat larger percentage of the whole ration is digested than when it is composed of fewer ingredients.

It is conceded by the majority of poultrymen that ground or soft food should form a part of the daily ration. As the digestive organs contain the least amount of food in the morning, it is desirable to feed the soft food at this time, for the reason that it will be digested and assimilated quicker than whole grain. A mixture of equal parts, by weight, of corn and oats ground, added to an equal weight of wheat bran and fine middlings, makes a good morning food if mixed with milk or water, thoroughly wet without being sloppy. If the mixture is inclined to be sticky the proportion of bran should be increased. A little linseed meal will improve the mixture, particularly for hens during the molting period, or for chickens when they are growing feathers. If prepared meat scrap or animal meal is to be fed it should be mixed with this soft food in proportion of about 1 pound to 25 hens. It will be necessary to feed this food in troughs to avoid soiling before it is consumed.

The grain ration should consist largely of whole wheat, some oats and cracked corn. This should be scattered in the litter which should always cover the floor of the poultry house. It is necessary to have the floor of the poultry house covered with a litter of some kind to insure cleanliness. Straw, chaff, buckwheat hulls, cut cornstalks, all make excellent litters. The object of scattering the grain in this litter is to give the fowls exercise. All breeds of fowls that are noted for egg production are active, nervous and like to be continually at work. How to keep them busy is a problem not easily solved. Feeding the grain as described will go a long way toward providing exercise. If the fowls are fed three times a day they should not be fed all they will eat at noon. Make them find every kernel. At night, just before going on the perches, they should have all they will eat up clean. At no time should mature fowls be fed more than they can eat. Keep them always active, always on the lookout for another kernel of grain.

GREEN FOOD.

While perhaps not strictly necessary for their existence, some kind of green food is necessary for the greatest production of eggs. Where fowls are kept in pens and yards throughout the year it is always best to supply some green food. The question how to supply the best food most cheaply is one that each individual must solve largely for himself. In a general way, however, it may be said that during the winter and early spring months mangel wurzels, if properly kept, may be fed to good advantage. The fowls relish them, and they are easily prepared.

Clover, during the early spring, is perhaps one of the cheapest and best foods. It is readily eaten when cut fine in a fodder cutter, and furnishes a considerable amount of nitrogen. If clover is frequently mowed, fresh food of this kind may be obtained nearly all summer, particularly if the season be a wet one. Alfalfa will also furnish an abundance green food. It must, however, be cut frequently, each cutting being made before the stocks become hard or woody.

A good quality of clover hay cut fine and steamed makes an excellent food for laying hens if mixed with the soft food.

Cabbages can be grown cheaply in many localities and make excellent green food so long as they can be kept fresh and crisp. Kale and beet leaves are equally as good and are readily eaten. Sweet apples are also suitable, and, in fact, almost any crisp, fresh green food can be fed with profit. The green food, in many instances, may be cut fine and fed with the soft food, but, as a rule, it is better to feed separately during the middle of the day, in such quantities that the fowls can have about all they can eat at one time.

GRIT.

It is necessary that fowls have access to some kind of grit if grain food is fed in any considerable quantities. During the summer months, when they have free access to the yards or runs, it will not be necessary to provide grit, providing the soil is at all gravelly. If, on the other hand, the soil is fine sand or clay, it will be necessary not only to provide grit during the winter months, but throughout the whole year.

Small pieces of crushed stone, flint or crockery ware will answer the purpose admirably. Many keep on hand constantly crushed granite in various sizes suitable for all kinds of domesticated fowls.

Crushed Oyster shells, to a large extent, will supply the necessary material for grinding their food and at the same time furnish lime for the eggshells. Chemical analysis and experiments, together with the reports from many practical poultrymen, show conclusively that the ordinary grain and green food supplied to laying hens do not contain enough lime for the formation of the egg shells. It will require several times as much lime as is ordinarily fed if good, strong egg-shells are to be produced. Crushed oyster shells will supply this necessary lime if kept continually before the fowls, trusting to them to eat the amount needed to supply lime rather than mixing the shells with food. The judgment of the fowl can be relied upon in this respect.

MEAT FOOD.

Where fowls are kept in confinement it will be necessary to supply some meat food. Finely cut fresh bone from the meat markets is one of the best, if not the best, kind of meat food for laying hens and young chickens. Unfortunately, it is not practicable for many poultrymen to depend wholly on this product, for the reason that it is often inconvenient or impossible to obtain, and when once secured it cannot be kept in warm weather without becoming tainted. Tainted bones should be rejected as unfit for food. Skim milk may be substituted wholly or in part for meat food without a decrease in egg production, provided the proper grain ration is given. The best meat food is dry animal meal, that is guaranteed wholesome. One to two pounds a day is sufficient for 25 hens.

FEEDING SMALL CHICKENS.

Chickens do not require food for the first twelve to thirty-six hours after hatching. One of the best foods that can be fed the first few days is stale bread soaked in milk. This should be crumbled fine and placed where the chickens have free access to it, and where they cannot step on it. One of the difficult problems for the amateur poultryman is to devise some means for feeding little chickens so that they can consume all of the food without soiling it. If placed on the floor of the brooder or the brooder run, the larger part of the food will be trampled upon and will soon become unfit to eat.

A simple and efficient feeding trough may be made by tacking a piece of tin about 3½ inches wide along the edge of a half-inch board so that the tin projects about an inch and a half on either side of the board, bending the tin so as to form a shallow trough, and fastening the board to blocks which raise it from 1 to 2 inches from the floor. (See Fig. 4.) The trough may be from 1 to 3 feet long. It is within easy reach of the chickens and so narrow that they cannot stand upon the edges. Food placed in such feeding troughs can be kept clean until wholly consumed.

Granulated oats (with the hulls removed) make an excellent food for young chickens. There is, perhaps, no better grain food for young chickens than oats prepared in this manner. It may be fed to good advantage after the second or third day in connection with the bread sopped in milk. A good practice is to keep it before them all the time.

The chickens should have free access to some kind of grit after the first day. Coarse sand makes an excellent grit for very young chickens. As they get a little older some coarse material must be provided.

Milk is an excellent food for these young fowls, but requires skill in feeding.

One of the great difficulties in rearing fowls is to carry young chickens through the first two weeks without bowel disorders. Too low temperature in the brooder, improper food and injudicious feeding, even if the right kinds of food are given, each plays an important part in producing these disorders. After the first ten days milk may be given more freely, perhaps, than during the earlier stages of the chick's existence. As the chick becomes a little older, more uncooked food may be fed. A mixture of fine middlings, wheat bran, a little cornmeal, and a little linseed meal mixed with milk makes a valuable food. Hard-boiled eggs may be fed from the beginning but, like milk, require more skill than the feeding of bread sopped in milk. On farms where screenings from the various grains become really a by-product, these form a cheap and efficient food for the little chickens. Wheat screenings, especially, form one of the best foods, particularly if they contain a considerable portion of good kernels that have been cracked in threshing. Then, too, screenings contain a number of weed seeds that have some feeding value and are relished by the fowls. They not only provide sustenance, but give variety, and this, in a measure, improves the general health.

Drinking fountains require close attention. Small chickens drink frequently and oftentimes with their beaks loaded with food, which is left, to a greater or less extent, in the water supply. As it is necessary to keep these fountains in a tolerably warm atmosphere they soon become tainted and emit a disagreeable odor. This condition must not be allowed to exist, for all the food and drink consumed by fowls should be wholesome. It has often been said that "cleanliness is next to godliness," and certain it is that cleanliness is next to success in poultry keeping. The drinking fountains must be kept clean. If automatic fountains are used great care must be exercised in keeping them clean and free from bad odors. Nothing less than frequent scalding with steam or hot water will answer the purpose. A cheap, efficient drinking fountain may be made of a tin can with a small hole in one end near the side of the can, under which is soldered a crescent-shaped piece of tin, forming a lip or a small receptacle for water. If the can is filled with water and then placed on its side, a small quantity of water will run out of the opening and remain in this crescent-shaped lip. As the chicks drink this water a quantity of air will pass into the opening and a little more water will flow out. This kind of fountain will keep before the chickens a small quantity of water at all times accessible. By exercising care and keeping the fountain thoroughly clean, satisfactory are easily obtained from this arrangement.

Brooder.

CHAPTER VII.

INCUBATORS AND BROODERS.

The modern improvement in incubators has made the rearing of fowls solely for egg production quite out of the question unless these machines are used. No experienced poultryman at the present time will undertake to rear fowls in large numbers for the production of eggs and depend on the hens that lay the eggs for incubation. The Mediterranean fowls cannot be depended upon for natural incubation. Artificial incubation must be resorted to if these fowls are to be reared in considerable numbers.

There are many kinds of excellent incubators on the market. As with many kinds of farm machinery, it is impossible to say that one particular kind is better than all others. Then, too, an incubator that would give very satisfactory results with one individual might prove to be quite inferior in the hands of another person. What is best for one is not necessarily best for another. It is advisable, before investing extensively in any make of incubator, to thoroughly understand the machine. If good results are obtained, then additional machines of the same kind should be purchased. Failures are recorded simply because the individual fails to thoroughly understand the ma chine he is trying to operate, or, in other words, fails to learn how to operate that particular machine to the best advantage. A successful poultryman must necessarily pay close attention to petty details. Not only is this necessary in caring for little chickens and mature fowls, but also in the care and management of incubators and brooders. The whole business is one of details. While incubators may vary considerably one from another, yet there are certain points to which all should conform. Some of these points are summed up in the following:

(1) They should be well made of well-seasoned lumber. The effort of manufacturers to meet a popular demand for cheap machines has placed on the market incubators that are not only cheaply made, but made of cheap and not thoroughly seasoned material.

(2) The incubator should be of easy operation. All its adjustments should be easily made and so arranged that the more delicate machinery is in plain

view of the operator. The machine should be automatic in operation. When supplied with the necessary heat it should control perfectly within certain limits the temperature of the egg chamber. The result is accomplished in various ways. The regulating force, whatever it may be, should be placed within the egg chamber so that the regulator may vary as the temperature in the egg chamber varies, irrespective of the changes of temperature of the room in which the incubator is placed. The regulator must be sensitive. The change of temperature which is necessary for the complete working of the regulator ought not to be more than 1 degree; that is, 1 degree above or below the desired temperature. It is better if the range of temperature can be reduced to one-half of one degree, thus making a total variation of 1 degree instead of 2 degrees.

It should not be inferred that a much wider variation than this will not give excellent results under otherwise favorable conditions, but, other things being equal, those machines which are most nearly automatic are to be preferred.

In addition to the foregoing requisites, a convenient appliance for turning the eggs, positive in its action, should accompany each incubator. The different machines may have very different appliances for accomplishing this result. Excellent results are obtained by the use of many machines now on the market when the operator of these machines is thoroughly interested. Poultrymen have for a term of years hatched in incubators over 80 per cent of all eggs put in the machine. It must not be inferred that it is an easy thing to do.

DIFFERENT KINDS OF INCUBATORS.

There are two general classes of incubators—those which supply the heat by hot water circulating in pipes and those which supply heat by hot air or radiating surfaces. Fig. 1 represents a 200-egg incubator heated by hot water. A tank of water is heated by a lamp, and this water circulates through pipes above the eggs. Fig. 2 shows a hot air incubator. A metal drum is heated by a lamp and the air is conveyed into the egg chamber. The hot air form is the cheaper, and for small experiments in artificial incubation is to be recommended. For larger operations the hot water variety is to be preferred. Some patterns may be supplied by water heated in a regular hot water tank, and a large number of incubators may be joined together and all be heated from the same fire.

Fig. 1

Fig. 2.

90

A double wall is necessary to secure uniform temperature. If a manufacturer has a double wall he will be sure to advertise it.

Hot water incubators should have an adequate system of ventilation, as fresh air is extremely important to good hatching.

GENERAL DIRECTIONS.

The incubator should be placed in a special room where there will be good ventilation and a uniform temperature outside of the incubator. There should be no drafts. A half basement is usually best, but a damp cellar is decidedly objectionable.

If the air is very dry, less ventilation is required to give the proper moisture; if the air is moist, more ventilation is needed to cause proper evaporation of the moisture in the egg.

Incubators should be visited at least twice a day, and the directions of the manufacturer of any particular make should be strictly followed, especially in regard to turning the eggs.

An amateur will do well to run his incubator a few days without eggs, in order to learn how to regulate the temperature. No incubator will be found to be strictly self-regulating. A record of the temperature should be kept at all times, both with a self-registering thermometer and an incubator thermometer.

The 200-egg size is usually to be preferred if one is going to have an incubator at all. For smaller hatches the natural method is nearly always to be preferred. If the egg-laying breeds cannot be made to set properly, a special yard of the Asiatic or sitting fowls may be kept expressly for incubation; but it will be necessary to keep the two varieties entirely apart the whole year round.

BROODERS.

If one resorts to artificial incubation it will be necessary to provide a brooder of some kind. It may be simple and quite inexpensive, or complex and costly. It is not necessary to expend very much money in the construction of an efficient brooder. It is necessary, however, to see that the brooder is capable of doing certain things. Some of these requisites are summed up in the following: It must be warm. The little chickens require a temperature of from 90 to 100 degrees the first few days, and at all times they should find it so warm in the brooder that they are not inclined to huddle together to keep warm. If the brooder is automatic, then the temperature may be kept quite even throughout the whole floor space. If, on the other hand, the brooder is heated from one side or from the top, and is not automatic, it will be best to construct it so that certain parts of the machine will be very warm, in fact, a little warmer than is necessary for the chickens, and some other part somewhat too cool. It does not take them long to learn just where the most comfortable position is. They may be trusted entirely to select the proper temperature if the brooder is of sufficient size so that it is never crowded. A brooder constructed on this plan will require less attention than almost any other. It may undergo a considerable variation in temperature without overheating or chilling the chickens.

The brooder should be easily cleaned and so constructed that all of the floor space can be readily seen. Inconvenient corners are objectionable in brooders; in fact, any corner is objectionable, but if brooders are constructed cheaply it is almost necessary to make more or less corners. If constructed of wood, circular ones are somewhat more expensive than square or rectangular ones. The floor must not only be kept clean, but dry.

As the chickens get a few days old, plenty of exercise must be provided. One objection to many of the brooders in the market is that the chickens are kept too closely confined and not allowed sufficient exercise. It will be a matter of surprise to many to learn how much exercise these little fellows require. With the young chicken, as with the athlete, strength is acquired by exercise, and above all other conditions of growth, strength is the one thing necessary in the young chicken.

KINDS OF BROODERS.

The brooders on the market are of the same two varieties as the incubators, those heated by hot water and those heated by hot air. The hot water ones are to be preferred when very large numbers of chicks are hatched; the hot air kinds will do very well for small lots, and are much cheaper.

Fig. 1 shows a brooder heated chiefly by hot water. It has a double floor, and the air space between the two doors is also heated. Brooders heated exclusively from the bottom are likely to give chicks leg weakness; but floors Should always be very dry and not cold. The double floor also economizes heat.

Fig. 1. Improved Brooder.

Fig. 2 shows another variety, with feeding pen. A home-made brooder may be constructed quite easily. A large box is turned bottom up so that the bottom will be about a foot from the ground. The top is inclosed with a tight board railing a foot high or more.

The top of the box forms the floor of the brooder. In this floor cut a hole, square or round, that can be covered by a tin pail with straight sides or a square tin box turned bottom up. In the bottom of the pail, now turned uppermost, cut a small hole and fix in it or over it by some means, such as by fireclay or cement, a metal tube rising six inches or so. The lamp (an ordinary kerosene lamp) is to be placed directly under this tin arrangement, so that the top of the lamp chimney comes about two inches under the opening or gas flue. A shield of tin should be placed between the top of the chimney and the flue, in order to throw the hot air from the lamp off on either side before it can go out at the flue. The tin pail makes a regular heating drum.

Fig. 2. Brooder and Feeding Yard.

A cover should now be provided that will have free air spaces around the sides for ventilation and light. It should be placed just above the tin pail and the small flue should run through it.

From the under side of this cover curtains of felt or flannel should be hung entirely around the heater, and from four to six inches apart. They should come down to within two and a half inches of the floor, leaving just space for the chicks to run under. These curtains make compartments of different temperatures, and the chick will choose the one it likes best. A thermometer should show the inside compartment kept at a temperature of 95 to 100 degrees.

CARE OF BROODERS.

The greatest care must be taken with any brooder to avoid chilling the chicks at night. For the first week the brooder may be kept inside, and then it may be placed out of doors during the day and brought in at night if the nights are chilly. It is well to provide a place for the chicks entirely separate from the other chickens so they may not be disturbed.

The great cardinal points in the management of brooders are:

1. That they be kept at a high uniform temperature, from 90 to 100 degrees for the first week, and somewhat lower after that.

2. They must have a good supply of fresh air.

3. They must be cleaned every day.

4. They must be dry and free from draughts.

If the chickens are seen to huddle together you may be sure the temperature is not right. Chickens kept properly warm will never huddle.

Of course an economical poultry raiser will see that the brooder can be heated economically, and gotten at easily and quickly, both to clean it and to examine the chicks.

CHAPTER VIII.

CAPONIZING.

First, does caponizing pay? The answer cannot be direct and unqualified. Except in the hands of experts it probably does not pay, as compared with other branches of the poultry industry. If skill fully handled, no doubt it does pay. It is more difficult to handle successfully than other branches.

Raising capons and broilers probably should be considered at the same time, and as a rule caponizing may be brought in to supplement the broiler business. Broilers require great care to have them in good condition early in the spring. It will he found impossible to bring some broilers to the proper condition when wanted. The poorer birds may be retained on the farm and prepared for the capon market.

Capons are castrated males. The castration is no more difficult to perform and no more inhuman than castration of any other variety of animal. After it is performed the fowls become more gentle and grow much, fatter and larger, the comb and wattles develop less rapidly, and so does the tail. The most important point is that they grow much larger. The flesh is highly prized, as it is tender. The birds fatten readily. The largest capons sell for the best price. The proper age for killing is one that does not exceed ten or eleven months. If kept longer, the meat becomes coarse and undesirable.

Early chickens will prove good for the broiler market, but later ones will prove more profitable for the capon market. The capons would then be available in the later winter months, whereas the same chickens as broilers would have to be killed in November or December when the market was overstocked with turkeys, etc.

Caponizing is easily performed. Full directions are given by sellers of caponizing instruments. An expert has been known to caponize 450 fowls in a day, and not lose one per cent.

All breeds are not suitable for caponizing. Black Langshans are considered the best, but Indian Games should be avoided. Plymouth Rocks do very well. Langshan and Plymouth Rock crosses are especially prized. Brahmas grow to large size and are good for capons.

Fowls should be operated upon when they attain the weight of about two pounds. A variation of half a pound either way will do no harm.

Chickens to be operated upon must be kept without food for twenty-four to forty-eight hours before the operation. As the abdominal cavity must be opened, it is important that the digestive organs should be free from food. The amount of blood is also lessened in this way, and bleeding is prevented to some extent. There are large arteries in the region to be operated upon, and the danger of cutting these is among the chief difficulties. Operations should be performed on a bright day, if possible. Artificial light and surgeons' head reflectors are sometimes used.

CHAPTER IX.

MARKETING EGGS AND POULTRY.

Nothing is so important in getting good prices as the method of handling eggs and poultry when sending to market. Looks go farther in the matter of poultry products than almost any other one item. The best egg or chicken in the world will sell for but half price if it does not look as it ought.

HANDLING EGGS.

Above all, eggs sent to market should be clean and free from any disfigurement. If the nests are kept clean, the eggs will usually be clean, but sometimes the first eggs of pullets are streaked with blood, and eggs will on occasion become soiled. A moist rag will usually clean them up without the expenditure of much time.

Uniformity in size is another prime necessity. A few large eggs in any lot will do an infinite amount of harm. It is better to keep the large eggs for home consumption. And when eggs are packed, all in a given case should be as nearly uniform, both in size and in color, as possible. White eggs beside brown, or speckled or streaked shells are always at a discount.

Fresh eggs are of course those most in demand; but it may be said that the greatest experts cannot tell the difference between an egg one day old and one four days old. In establishing a reputation for eggs strictly fresh, a poultryman will at first probably mark all his eggs and give a guaranty with each lot that they are strictly fresh. The marking may be quickly done with a small rubber stamp.

Eggs are usually sent to market in crates with pasteboard partitions on all four sides of the egg. Many commission men, however, prefer eggs packed in barrels in dry, fine, clean straw (wheat or rye). Seventy dozen eggs may be packed in a barrel.

FATTENING FOR MARKET.

In fattening fowls, care should be taken to give young fowls some exercise in order to keep them in a healthy and vigorous condition. Old fowls require little or no exercise. Especially should little exercise be allowed for a few weeks just before killing, if a choice quality of meat is desired. Close confinement improves the quality of the meat.

Pure air and an abundance of soft food are the chief requirements. For quick fattening the ration should consist largely of corn. But a variety of food, including wheat, barley, buckwheat and oats, will serve to maintain a good appetite longer than any single food. After fowls have been kept some time on soft food, hard food cannot be given them, since they will be unable at once to grind it properly. Young fowls of fine quality are often fed from 10 to 20 per cent of animal meal, but no one would think of feeding such a proportion to older fowls.

The sexes should always be separated before the fattening period begins. It is desirable to have the fowls all of a size as far as possible. Then each gets its own share of the food. If cocks which are being fattened have a disposition to fight, they may be placed in coops with slat bottoms, which will not give them sufficient foothold to stand up to each other.

DRESSING AND SHIPPING.

A considerable portion of the dressed poultry consigned to the commission houses in large cities brings to the producer a much smaller profit than it would had the same poultry been dressed and packed for shipment with greater skill. It is of prime importance that the poultry products be placed on the market in a condition that will make them appear as inviting as possible. Proper feeding for two or three weeks before the fowls are slaughtered will improve their color materially. In most of the American markets fat fowls with a yellow skin bring the highest price. This condition may be secured most cheaply by feeding a grain ration composed largely of corn for two or three weeks before the fowls are slaughtered. Of the more common grain foods there is none that excels corn for this purpose.

The commission men and shippers, who study in detail dressing and packing, state that uniformly fine quality will soon acquire a reputation among buyers. The shipper should always be careful to have the product look as neat as possible. In some of the large cities ordinances prohibit the sale of dressed poultry with food in their crops. In a few instances the sale of live poultry in coops which contain food is also prohibited. In all cases it is best to withhold food from twelve to twenty-four hours before killing, but the fowls should have plenty of water during this time, that they may be able to digest and assimilate food already consumed. All fowls should be killed by cutting through the roof of the mouth and allowing them to bleed to death. In all operations of dressing avoid cutting or bruising the skin or breaking bones. Care is required in the case of the heavy fowls in picking and handling to prevent bruising the skin. In packing fowls use neat, clean, and as light packages as will carry safely. Boxes or barrels holding about two hundred pounds meet these requirements best; boxes are better for turkeys and geese and barrels for chickens. Barrels may be used, however, for dry shipment as well as for hot weather shipment when the fowls are to be packed in ice.

In shipping live poultry the coop should be high enough to allow the fowls to stand upright without bending their legs. When large coops are used there should be partitions, so that if the coops are tipped all of the fowls are not thrown to one side. They should have plenty of room in the coop. If possible put only one kind in a coop or in one division of a coop.

CHAPTER X.

Duck, Goose, and Turkey Hatching and Raising.

HOW TO MANAGE RAPID GROWTH OF DUCKS.

Ducks are profitable if sold as soon as they reach four pounds weight, or five at the highest, as they will retain rapid growth and increase for all the food they may consume up to that age. After that time they do not pay except to keep a few, unless they have a pond and grass run. If raised under hens, keep the hens and young ducks in little coops and runs, away from water. In fact, until the ducks are feathered, they should be given drinking water in a manner only to allow of their getting their beaks wet, for contrary to the old saying that "wet weather is splendid for young ducks," nothing is so fatal to them as dampness. Very cold drinking water will cause them to have cramps, hence it should be tepid.

The duck is a very hardy fowl, much more so than any other class; the Pekin being the most popular and probably the best example. The eggs are usually fertile, hatch well in the incubator, or under fowls, and a large per cent can be raised to maturity, the estimated mortality having been placed as low as two per cent.

The growth of the duck is more rapid than that of any other fowl known, thus making a broiler early in the season, when prices are good. In addition to this the feathers are always in demand and they can be plucked every six or eight weeks from those that are carried beyond the broiler age. The most reliable authority will always advise using only those ducks that were hatched in early March or April, for breeding purposes, not any later, for the former (those hatched in early March) will give the best service, which requires the setting of eggs about the first of February, as it takes four weeks to hatch ducks. Now the question arises, how are these early breeders and broilers to be had, for at this season of the year there are not many ducks or hens ready to set, and if there are, the result of the hatch is very unsatisfactory, as the fowl must of necessity leave the nest, often absenting itself much longer than the delicate condition of the developing duckling is able to endure, with the

101

result either of a dead embryo in the shell, or a very weak and probably crippled fowl.

Just here is where the Incubator and Brooder, that boon to poultrymen, are available; with them you are prepared to have a hatch come off just when you want it, thus enabling you to take advantage of what is necessary to make this business a profitable one. The eggs used for hatching should be clean, although duck eggs ought not to be washed, or if so, as little as possible, but this will not be necessary if the poultry house is kept clean. Dirty eggs, besides not looking well in a hatcher, are harder to test. The oily, greasy covering on the outside shell should remain, for it has been demonstrated the eggs hatch better if it is not removed.

The eggs should be tested after they have been in the machine about seven days, and the unfertile ones removed. When the hatch is finished and the ducklings are thoroughly dry, they should be removed to the brooder, where their first food should consist of a little corn meal and stale bread, moistened and well mixed so that it will crumble.

Feed them after they are 24 hours old on a mixture of mashed potatoes, which may be thickened with ground grain (composed of equal parts of corn meal, ground oats and middlings), and give them all the milk they can drink. Scald all the food the first two weeks. After they are three days old give them meat, chopped fine (or ground meat), mixed in their food three times a week. Chopped grass, cabbage, vegetable tops, clover hay, chopped and steeped in water, or any kind of green food may be given liberally. After the second week cooked turnips and ground grain will answer, with a little ground meat. Feed four times a day until they go to market.

When raised in brooders feed them in little troughs, to avoid fouling the food. They require plenty of heat in the brooder at first, but after they are four weeks old can do well without it. Give them plenty of drinking water always, and let it be clean.

From January 1 to May 1 is only 18 weeks, and as the ducks must lay enough for hatching, a little time will be lost in that direction. Then the ducks may have to be dieted to get them into proper condition, as many make the mistake of feeding them too much grain, thus having them too fat. If this is the case the best course to pursue is to feed them only once a day for a week, late in the evening, on some bulky food, such as cooked turnips with a small allowance of bran. Plenty of water should be supplied, however, and skimmed milk may be freely given. At the end of the week the ducks should be fed on cooked turnips, with more bran, and some kind of animal food, such as ground meat, or fresh meat from the butcher. This may be given twice a day, but if they begin to lay feed them three times a day. It is best to endeavor to separate the layers from the non-layers until they are laying regularly. Always mix their soft food with skimmed milk, if it is plentiful, and use ground oats, bran, or chop in preference to corn or corn meal, in order to avoid making them too fat.

The ducklings may be allowed in the yards as soon as they feather, and even very young ones may go out on warm, clear days. They are sent to market "dressed," truly express it "undressed," for they must be dry picked, and all the pin feathers removed. In all other respects they are treated in the same way as broilers. No one who has ever dressed a duck will worry for a repetition of the job, and no one knows how many feathers are on a duck until he attempts to get them off. With ducklings it is even more difficult; owing to the large proportion of pin feathers. After a duck is supposed to be picked clean it may be picked over again half a dozen times, for they will still have a fuzzy appearance. The dressing of the ducklings is very disagreeable, but if you get your ducks laying early and manage to have a large number in market for the high prices, you will be amply rewarded for all your trouble.

They are subject to but few diseases. Cramps occur from cold water. Leg weakness comes from damp quarters at night. Apoplexy attacks grown ducks when they are very fat, and they are also subject to vertigo. If attacked by the large, gray body lice on the heads, they will appear apparently well, and suddenly turn over on their backs and die. The floor upon which they sleep must be of boards, and should be kept very clean and dry. As we stated, dampness is fatal to young ducks.

A duck of the improved breeds will lay from 120 to 160 eggs per year, and usually begins in February. If kept in the house until about 8 o'clock in the morning, they will lay in the house, as they lay early in the morning, but if turned out too soon, they sometimes deposit their eggs in other places, and even on ponds. One drake to six ducks will be sufficient, and if young females are used; it is best to have a two year old drake, though sometimes the eggs hatch very well from parents of the same age on both sides and less than a year old, but they should not be kept too fat, or the eggs will not hatch well.

It requires, on an average, three months for a chick to reach two pounds, while a duck arrives at that weight (averaging a number) in less than half that time, and is ready for market (weighing three pounds) in seven weeks, thus giving really a large profit in the summer months, though prices rapidly decline after July 1st.

It will be noticed that ducks gain faster at some times than at others, which difference is due to warm or damp weather, as the case may be. After the fourth week they should gain from 8 to 12 ounces each per week, though we have known them to gain 14 ounces in one week. Everything depends upon the food and care bestowed.

The best breed, as before stated, for the purpose is the Pekin. It will thrive without ponds, and is hardy and easily raised. The yards must be kept free from the filth. Ducks need water with which to wash their bills when eating, or their nostrils will become clogged and suffocation will result. Never use very cold water for young ducks, as it causes cramps, but have it lukewarm. Leaves, cut straw, and hay litter may be used in their quarters. The most frequent cause of disease and death of young ones is filth, but over-feeding causes leg weakness and carries off the old ones.

CARE OF GEESE.

Geese require very different surroundings and care from ducks. Geese need water much more than ducks, and they need a much larger range. They can also be raised only in comparatively small flocks. There are few very large geese raisers.

The best place for geese is a pasture where there are many springs, or springy and marshy ground. If they have a good place they seem to require but little attention. They live chiefly on grass and worms and insects.

Young goslings begin life by eating grass. Moistened corn meal is a suitable food. Of course grit must be provided. Coarse sand will do. A mixture of bran, middlings, and corn meal, cooked with vegetables, is to be recommended. Animal meal should not be forgotten for the, young; and also for older geese when confined in winter.

When fattening geese, confine them closely, to prevent exercise, and avoid any disturbance.

Simple houses which offer a suitable shelter from storm, etc., are all geese need most of the year. But it is very important that they have a dry place to sleep. Nothing kills young goslings so quickly as moist houses. Yet geese need ponds for swimming and bathing much more than ducks do.

TURKEYS.

A wide range is indispensable in raising turkeys. A small village lot or a small farm will not do, for if confined in any way the turkeys do not thrive.

At the laying season turkey hens will usually try to hide their nests, and at this season may be somewhat confined. The eggs should be removed as laid, and the first put under a setting hen. A hen turkey will usually lay more eggs than she can cover well, and so she should be allowed to sit only on the ones last laid. Turkey eggs, like those of geese, hatch in about twenty-eight days. Hatching turkey eggs by incubation has not proved satisfactory.

Turkeys are best bred from old stock, and when one finds a good mother and a good breeding gobbler, they should be kept to breed from as long as they are good for anything.

"Young turkeys should not be out in heavy showers until their backs are well covered with feathers. If they get wet they may die from chill unless put in a warm room to dry. Black or red pepper or ginger in food or drinking water aid them to overcome a chill, and are of great value on cold or damp days, and are a preventive of bowel troubles in both old and young turkeys."

Above all, young turkeys must have a dry place to sleep, and dry, porous soil is far better than heavy soil even if it is drained. At first the young turkeys should never be allowed to wander in wet grass. Hard-boiled eggs finely chopped make an excellent first food. Stale bread dipped in milk is also good to begin with. No food should be sloppy. Later the best chicken feed is usually good also for turkeys. In feeding corn later in the season, old corn is preferred to new.

MAMMOTH BRONZE TURKEYS.

The Mammoth Bronze is the king of all turkeys. Plumage of male, on back and breast, is a brilliant bronze hue, which glistens in the sunlight like burnished gold. The wing coverts are a beautiful bronze, the feathers terminating in a wide bronzy band across the wings when folded, and separated from the primaries by a glossy black ribbon-like mark, formed by the ends of the coverts.

Tail. Each feather is irregularly penciled with narrow bands of light brown, and ending in a broad black band, with a wide edging of a dull white or gray. In the female the entire plumage is similar to that of the male, but the colors are not so brilliant or clearly defined, and the edging of the feathers is generally a dull white or gray.

Mammoth Bronze Turkeys.

The Mammoth Bronze is the hardiest of all turkeys, and the most extensively raised of any breed. They are good layers, many claiming them to lay over 100 eggs in one season. However, there are exceptions in all things, but it is no unusual occurrence for a turkey hen to lay 50 eggs during hatching season, say from April 1 to July 1. Most turkeys do not lay after July 1.

Standard weight for toms, 32 pounds; hens, 22 pounds.

ROUEN DUCKS.

Rouen Ducks.

The Rouen duck is considered one of the hardiest of the duck family. The head of the drake is a lustrous green, bill is a greenish yellow. The neck has a distinct white ring on the lower part, which should not quite meet at the back. The back is an ash gray mixed with green, with lustrous green on the lower part, the saddle coverts being streaked with brown lines. The breast is a purplish brown. Under part of the body is gray, being lighter behind with black under the tail. The wings are grayish brown, mixed with green, with a broad ribbon of purple, with green and blue tinges, edged with white. Tail is ashy white, the outer web in old birds edged with white. Legs are orange with a brownish tinge. The duck is a deep brown on the head, with two light brown stripes on each side, a dark orange bill, having a bean on the tip and a dark blotch on the upper part. The neck is a light brown with no penciling. The back is light brown marked with green; breast, dark brown penciled; under

107

part of body and sides grayish brown and penciled with darker brown; wings, grayish brown mixed with green, and having the purple ribbon bars across them; tail brown; legs orange. In size they are the same as the Pekins, are very hardy, and splendid layers. Ducklings grow very rapidly and are full fledged at about 10 weeks if well cared for.

IMPERIAL PEKIN DUCK.

Imperial Pekin Ducks.

The Imperial Pekin Duck is the most popular of all ducks and is most extensively raised. They are a large white duck, in many instances they are a rich creamy white, but this rich color does not show at all times, only when they are in full feather and in prime condition. They are very hardy, ducklings at the age of 8 weeks weighing sometimes over 4 pounds and in full feather. They are splendid layers, having known them to lay as high as 75 eggs in succession, but they lay best when in small flocks. If kept in large flocks they do not lay so well. Their eggs usually hatch well and ducklings are strong and vigorous. It is erroneous to say ducks and geese must have running water to thrive well. We have seen ducks and geese raised (and they were remarkably thrifty) with only plenty of water to drink and no water to swim in at all. But they enjoy themselves much more if they are provided with water to swim in, or at least to wash themselves in. Pekin ducks usually weigh from 7 to 9 pounds each.

EMDEN GEESE.

Emden Geese.

The Emdens are an excellent breed of geese. In size they are about the same as the Toulouse, and in laying qualities they are about the same, but in color they are a pure white. This is one point in favor of the Emden, as the pure white feathers are considered much nicer than gray feathers. The expense of keeping them is very small where a grass run and plenty of water are provided for them. The above illustration gives a good idea of their general appearance. They are hardy, and goslings grow very fast with good care for the first two or three weeks, from which time they will help themselves if plenty of water and grass are provided for them.

TOULOUSE GEESE.

Toulouse Geese.

The Toulouse Geese are a purely English breed. Both male and female are very massive in proportion. The bill and feet are dark orange color, head, neck and back a dark gray, breast light gray, but descending lighter, till beyond the legs to the tail they are pure white. The combination of colors presents a very attractive appearance.

Both male and female are uniform in color, being alike to a feather. They live to a great old age; some have reported them living and doing well at the age of 30 years. Goose raising is very profitable, as they need no grain in the summer, when they can have grass to feed on, and are small feeders in winter. While the expense is so light to keep them, one can also pick their feathers four times in one season, making about 2 pounds of feathers from one goose, worth $1 to $1.50. With the young one can raise they become very profitable. The weight of an adult goose is 23 pounds, gander 25 pounds. Old stock produces stronger goslings than young ones do.

CHAPTER XI.

DISEASES: HOW TO PREVENT THEM AND HOW TO CURE THEM.

GIVE THE BEST OF CARE.

One reason that so many do not have good results in raising chickens as they desire, is that they undertake to raise more than they are fixed to take care and the result is the young birds often die, and others do not develop into good birds, being under size, and lacking in plumage. Take chicks hatched from a pen of Black Langshans or Brown Leghorns, overcrowd, and do not feed them with plenty of good, nourishing food, and the result will always be, that white feathers will show in wings, or some other defect equally objectionable; while chicks hatched from the same yards, and given the care and food necessary to keep up a healthy growth, a good share of prize winners will be the result. This is not a theory, but has been demonstrated a great many times by breeders who have been careful observers.

Have the boys and girls keep a systematic account of the poultry business, so you will be able to determine the profits. It will at the same time be teaching them lessons in business that will be valuable to them.

Burn corn or wheat until in charcoal state, and give to the poultry. It shows in bright red combs; and healthy appearance, is an excellent corrector when birds are ailing, better than drugs, and cheap. Keep fresh, clean water daily in vessels in yards; keep all roosting places cleaned up often and use coal oil freely in building and on roosts; gather leaves from the woods and have a good supply to use through winter in yards and sheds. Sow grain among them and the fowls will scratch after the feed, getting exercise and health. Keep dogs and small children out of the chicken yards. They scare the birds and often do more damage than a dozen rats. Look after the birds yourselves, not trusting to the hired man or boy for everything. Your presence is of use, and one interested can see what needs doing. Keep a supply of gravel on grits always in the yards when poultry are kept in small quarters; when they run at large they find such things themselves. Havel fresh ground bone from the butcher's fed at least twice a week; it helps in growth of young birds and laying plenty of eggs in mature ones. Give a variety; don't feed one thing long at a time.

HOW AND WHAT TO FEED.

For ten hens one quart of grain is sufficient for a day's ration. Especially for the smaller breeds.

However, Brahmas or Cochins if they do not have access to green food, will require about one quart to five heirs during laying season.

All fowls relish a cooked breakfast or dinner, potatoes, turnips, carrots, cabbage, or anything in the way of scraps from the table, equal parts of corn and oats ground together, and middlings boiled or steamed, if served while warm makes a good breakfast for laying hens. Another good food for the morning ration, is corn chop, ten pounds; barley chop, ten pounds; middlings, five pounds, and bran, five pounds. Mix all together, then scald with just enough boiling water to make a thick mash; add sufficient water or milk to cool so that it can be readily eaten by the fowls.

For an evening food we usually give whole grain, either wheat, barley or oats. Corn is not good food in summer as it does not contain the required ingredients to produce an abundant supply of eggs, and is productive of too much heat. Never feed corn during the laying season except that part contained in the soft mash which is given to the fowls in the morning. Another important factor in egg production is green food. In order to have fowls do best, they must be supplied with yards large enough to furnish an abundant supply of grass. However, if there is no room for this accommodation they must be supplied some other way, which may be done by mowing grass and cutting it fine by running it through a clover cutter. Fowls fed with grass thus prepared will do quite well, and they eat it very readily. If they are supplied with sufficient cut grass in small yards they will lay as many eggs and thrive just as well as those in larger yards where they can eat grass at will.

WHY DO HENS STOP LAYING?

The most profitable hens lay at seasons of the year when eggs are scarce and dear. The thing to do, then, is to provide conditions which will be conducive to egg production. Laying hens depend less upon the seasons of the year than they do upon the weather. That being the case all poultry keepers should provide comfortable quarters for the hens in both winter and summer says "The Western Stockman."

Why is it that the hens will be in good condition and as soon as the first cold snap comes every one of them will stop laying? It is a matter that is worthy the attention of poultrymen or philosophers. It is a serious thing when eggs are selling at three cents apiece, to have the hens suddenly cease laying when they should be filling the egg basket. It seems hard for the hens, after laying as regular as clock-work during the summer; when eggs are low, to shut down just when there is the heaviest demand.

But there is a cause for it. It is not because of lack of food, as the cessation of egg production may happen in a single day. It is not due to disease, for the hens may be healthy. The cause is lack of warmth. While the heat of the body comes from the food, yet the cold is so intense that digestion is not sufficiently rapid to create the heat necessary to protect the bird from the cold. Egg production ceases because nature's first effort will be to protect the bird before it is permitted to do extra work in production. What is the remedy? It is simply to guard against the loss of animal heat. This is done by keeping the cold winds away, by providing sheltered and sunny places for the hens, by feeding warm food and warm water. No ventilators, cracks or openings are needed to let in the cold air. If you wish the hens to lay as they do in summer they must have summer conditions.

A FEW HINTS.

Very fat fowls are poor breeders, and are mere liable to lay soft-shelled eggs. Give the sense intended for breeders sweet, nourishing food and keep them in motion, but be careful not to overfeed with corn in any form.

The wise poultryman sells as soon after maturity as a good price can be obtained; he has a definite purpose when he feeds, and gives such treatment as is essential to the procuring of the best growth and condition. Only business methods pay in any business.

England paid $22,000,000 for poultry and eggs last year from other countries. This is true not withstanding it raised more of its own than ever before. In fact, the consumption of these products of the farm is greatly on the increase the world over.

With the use of very simple tools it is easy to keep the hen-house in good shape. There should be a large bucket of light weight, a garden rake with which to gather up the feathers and litter from the floor, a piece of thin board for cleaning off the roost board, and a shovel.

It is thought by many that symmetry and the quality of the offspring can be largely influenced by care in the selection of the eggs for hatching; that those from defective fowls of any sort should not be used and that eggs uniform in color and size give the best results.

Poultry requires feed, and America has the feed in the greatest abundance, and can produce poultry to feed the world. With our improved breeds, incubators, modern poultry houses and poultry palace cars, there is no limit to the production of table poultry and eggs for the city market.

A few hens in a single enclosure will pay a profit invariably, if properly cared for, but a lot of neglected, crowded fowls will seldom return the food they consume. It does not follow that you can get $200 net profit from 100 hens just because 10 hens brought you $20.

It has been thought that poultry could not be successfully raised in large numbers, but the science of breeding and poultry management has developed the way with the incubator and the separating of the flock into small numbers. Improved breeds are kept by thousands by some of the larger poultry men near the cities.

The fattening of fowls for the market is a thing which can be done quickly if the appetite is properly tempted. While thus crowding them it pays to give them clean food and often rather than to keep a supply before them all the time. Do not sell your surplus roosters for a song, but shut them up in a quiet place, and feed them heartily for a few weeks.

Prompt action may prevent the spread of disease among the flock and great loss to the owner. A very sick hen is not worth doctoring, unless she is of especial value, and the sooner her head comes off the better for all concerned.

If cockerels are separated from the hens at an early age they may be grown to a large size and still be excellent food for the table; at least so say those who are opponents of caponizing. They claim that a considerable proportion of the birds sold in the Parisian markets are but cockerels which have never been allowed to run with the hens.

When space is limited be all the more careful about keeping the quarters clean, especially if the chickens cannot get out much, and do not overstock. It is not too much even if there is a cleaning made under the roosts every morning, at the time the stable is cared for.

ROUP IN ITS VARIOUS STAGES.

One of the most dreaded diseases among poultry is that of Roup, which usually begins with a cold. All fowls are subject to colds, as well as humanity, and should have the same attention that we would give ourselves; for should we neglect to apply a remedy when we take cold the result might prove quite serious. The same will be applicable in case your fowls take cold, which may be brought about in numerous ways, viz.: roosting in damp quarters, cold draughts of air passing over them in their sleeping apartments, sleeping in brood coops on the ground where they are packed so close as to smother some during the night, and those not suffocated are over-heated so that when exposed to the cold air in the early morning a severe cold is the result, and if a remedy is not speedily applied and the cause removed, Roup will invariably follow, which of all poultry diseases is the most obstinate, sickening, and difficult to cope with, and if necessary precautions are not taken in the start to arrest the disease, it will run through the entire flock and leave nothing but death and destruction in its path. In our opinion Roup is more to be feared by poultrymen than the deadly disease, Cholera.

Symptoms of Roup may be described thus: Fowls begin coughing, sneezing, and sometimes their breathing is heavy, accompanied by a wheezing sound. Eyes become inflamed, heads swell and they have a watery discharge from their nostrils which sometimes has quite an offensive odor; they are drinking almost continually if they have access to water, which is an indicative of their being feverish. As the disease advances the head becomes inflamed, swelling on one or both sides, frequently obstructing their sight, the eye sometimes being entirely destroyed. It may be noticed when fowls are affected with this disease they have splendid appetites and eat until the last, provided they are not internally affected, in which case they are stupid and a discoloration of their excrement may be noticeable, which is much the same as that of fowls affected with Cholera.

Cure for Roup.—When fowls are in the advanced stages of the disease, the best remedy is the hatchet, as they can seldom be cured, although in the early stages they may be cured by taking a small spring-bottom oil can and injecting in their nostrils and roof of their mouths a little kerosene oil; if heads are swelled, anoint the parts swollen with sweet oil and alcohol, equal parts each day. Add some good condition powder to their morning mash. Put about one-half teaspoonful of aconite to each quart of their drinking water. Keep them in good, dry, comfortable quarters, with an abundance of sunshine in their room, and it should be well littered with straw or leaves, which must be changed frequently. Their drinking vessels should be cleansed with boiling water. The utensils in which they are fed their morning mash should also be cleansed with boiling water, as this is absolutely necessary to accomplish a speedy cure; not forgetting to remove all sick fowls from those not affected, to prevent spreading of the disease.

A subscriber to the *Poultry Keeper* gives another remedy which he says cures the Roup every time: When the chick first shows symptoms of Roup, open its mouth and with a small glass syringe insert into the throat as far as possible a little finely pulverized alum. If the disease is in an advanced stage, and the head begins to swell, anoint the swollen parts with common vaseline, also insert some of the vaseline into the nostrils with a small feather. Feed on a liberal supply of bread and milk well seasoned with pepper; to one teacupful of soft food, such as bran or oats, mix one teaspoonful of castor oil. Do not neglect to place those affected in a dry, warm place.

Kanker is another disease which may be classed with Roup. This disease is quite offensive, but it is not difficult to cure. It forms in different parts of the mouth, but mostly at the base of the tongue, or in the windpipe. Sometimes people call this disease diphtheria, which may be the proper name for it, as it resembles that disease very much. However, diphtheria among children is sometimes very obstinate and difficult to eradicate, while among fowls it yields to proper treatment quite readily. While some may call this disease among fowls diphtheria, I should prefer to speak of it as kanker, which forms in a white or yellow leathery substance in the mouth, as before stated. Remove this substance with some kind of instrument, which may be a stick, nail or anything with which you can perform the work, then apply sub sulphate of iron (powdered form) to the parts affected, which may be obtained at any drug store, and ten cents worth will be sufficient to cure many cases. The above is a positive remedy, and a change for the better may be noticed after the first application, yet it should be repeated until an entire cure is effected, which usually is only a matter of a few days.

GAPES.

Old fowls are never afflicted with the gapes. The disease is found only among chicks, and is caused by a worm or worms which infest the trachae. When once noticed on the premises it can never be entirely eradicated. It appears to be in the soil, and chicks each year will be subject to the gapes more or less after the place has once become contaminated therewith. Gapes among chickens may be cured by the use of a horse hair. Twist one together so as to form a small loop at one end; insert this end down the wind pipe and if you turn it around several times, the worms get caught in the loop and can be drawn out. Here is another remedy which, it is claimed, never fails to relieve the chicks of the gapes, and with proper care you will not lose a bird. Take a tight box about three feet long, one foot high, and one foot wide; place a partition crosswise about twelve inches from one end, made of lath or wire screen. Then place a brick or stone on the floor in small end of box; after this take a piece of iron and heat it red hot. While the iron is heating catch the chicks that have the gapes and place them in the larger end of the box. Take the red hot iron and place it on the stone or brick and pour a spoonful of carbolic acid on it. Close the box for a minute or two, then open and stir the chicks around so that they all can inhale some of the gas, which will kill the gape worm. If some of the chicks are overcome, lay them out and they will soon revive again. Do not leave them in the box too long or the gas will suffocate them. The first application usually cures, but should there be any that has not been cured with the first dose, repeat the second time, and it will never fail to cure them.

SCALY LEGS

Is usually caused by the chicks or fowls sleeping in filthy quarters. It is also caused by a small parasite which works underneath the scale of the leg. I have seen fowls with scaly legs that were twice their natural size. If the legs of each

117

fowl were anointed once each month with equal parts of sweet oil, kerosene oil and alcohol they would never become scaly, but would remain in a fine, healthy condition. A good remedy is lard and kerosene oil, equal parts; add enough pulverized sulphur to make a paste, then apply this to the legs and bandage them, leaving the bandage on for a week. If at this time the scales are not all healed off, repeat the application of the same ointment, as it is a sure cure. The bandage may be sewed on so that it cannot be scratched off by the patient.

DYSENTERY.

Dysentery in chicks is invariably brought on by irregular heat. If quite young chicks get chilled, bowel complaint will be the result. If overheated the same disease will follow, which is fatal in most instances; at least it retards their growth. Never allow chicks to get chilled or overheated if you wish them to do well. Usually during the warm summer months the most difficulty is experienced in this line, owing to the warm days and cold nights we often have at this time of the year. Cure: To a pint of soft food add a tablespoonful of finely ground raw bone, which should be fed at least three times a week to the healthy chicks as well as to those affected. Boil two ounces of ginger and one ounce of copperas in a gallon of water. Moisten the food with this fluid, but avoid feeding corn in any form when chicks have the dysentery.

LEG WEAKNESS

Is found chiefly among chicks raised in a brooder warmed by under heat. Or it is sometimes brought about by high feeding; in this case their bodies grow too fast for the strength of their legs. If the bottom of the brooder is slightly warmed it will do no harm, but the most of the heat should come from above, and then you will scarcely be troubled with leg weakness among chicks. Those that have leg weakness will in course of time come out all right, without the aid of medicine, and they usually make the finest specimens, as only the most vigorous chicks become affected. Feed finely ground raw bone in the soft food daily, which will strengthen their legs and will be the means of their rapid recovery.

BUMBLE FOOT.

This disease is caused by fowls running on hard, dry ground when confined in small runs, as it is seldom found among fowls that have the run of the farm where they get the wet grass in the morning.

Symptoms—They become lame, with inflammation and swelling in the foot. The bottom of the foot is hard and has a scab which should be removed and the core pressed out; but should it be fast, take a knife and make a deep incision in bottom of the foot and take the core out in pieces. Sometimes they swell between the toes; then the incision should be made where the swelling is. After you have made the incision make a poultice of linseed meal and place the foot therein, which will draw the pus all out. However, should any core form in the wound while under treatment, be careful to remove it. Apply a fresh poultice every day, and nine times out of ten you can effect a cure.

CHOLERA

Is more prevalent in warm than in cold climates. It is a bacterial disease and is highly contagious for the simple reason that the bacteria germs are ejected with the excrement and the healthiest and most robust succumb to its ravages alike with those that are more delicate. Investigation by the government officials shows that the first symptoms of chicken cholera is, in the great majority of cases, a yellow coloration of that part of the excrement which is secreted by the kidneys and which is normally of a pure white. This yellow coloring matter appears while the excrement is yet solid, while the patient presents a perfectly normal appearance and the appetite is good, before there is any elevation of the temperature. In some cases the first symptom is a diarrhœa, the excrement being passed freely, and after a day or two it becomes a dark green in color. The comb is pale or bloodless and sometimes of a dark purple or blue.

The duration of the disease varies greatly; sometimes the bird dies within ten hours of the first attack of the disease, and again they will sometimes linger for several days.

There are numerous remedies for the cure of chicken cholera. In the first place isolation is necessary; give them a warm, dry and comfortable house. Disinfect the premises thoroughly with a solution of eight ounces of sulphuric acid to two gallons of water; sprinkle the ground and everything in the house thoroughly with the disinfectant; remove all the droppings from the house and away from the healthy fowls. To each gallon of drinking water add a teaspoonful of carbolic acid. This is also a good disinfectant and will act as a preventative. The following recipe is one that will be found efficacious in the cure of the disease:

Isolate those affected, and give each a pellet about the size of a grain of corn or a pea, three times a day, made from the following powder (use a little flour and water to make the pellets):

2 oz. Capsicum,	1 oz. pulverized Rhubarb
2 oz. pulverized Asafetid	6 oz. Spanish Brown,
4 oz. of Carbonate of Iron	2 oz. Sulphur.

As a preventive, add a tablespoonful of the above powder to the soft food for every ten or twelve fowls, twice a week.

CROP BOUND.

There is a disease which exists among fowls and is probably the commonest of all crop diseases. It is caused by their eating more than they can digest. It is easily detected, as a fowl with a full crop, if in a good healthy condition, will carry it up firmly. But when crop bound, the crop will be loose and hang down like a bag, which may only be a ball of hard, coarse food that resists the force of nature in digesting. If not left too long it can be removed by pouring some warm sweet milk down the patient's throat into the crop. Then work the hard substance until it becomes soft, which in some instances may take an hour. Repeat this daily until a cure is effected. Feed only bread sopped in milk until the patient has entirely recovered. If the case is one of long standing it will probably require the use of a knife. Make an incision lengthwise near the top of the crop and remove the contents through the incision. However the contents is sometimes so hard that it will have to be broken up before it can be removed. After the sour food has all been taken out, then take a needle and some silk thread and sew the orifice up—first the crop and then the skin. The patient should then be fed on bread dipped in milk or some soft food that may be readily digested; continue this food for five or six days after the operation has been performed. No water should be given as long as the feeding of the soft food is continued.

LICE.

Cleanliness will usually prevent the appearance of lice. They are first found on the poorer and weaker fowls, and it is believed that they must be introduced by an infected fowl. This is a reason against buying grown fowls rather than raising from eggs. It has been estimated that a single pair of lice in three months will produce 100,000.

A few drops of sweet oil or lard on the head, wings and throat of little chickens will prove best. Older fowls should be allowed nature's remedy—dust baths. Powdered sulphur or insect powder dusted into the feathers is good. Some put fowls in tight boxes, with heads protruding, and fumigate with sulphur fumes for a few minutes. This is said to do no harm and kill all pests. If the poultry house is infected it should be thoroughly cleansed—whitewash, sprayed chlorides, or an emulsion of kerosene oil (if spraying is done thoroughly) being recommended for this purpose.

MITES.

These pests are very different from lice and live by sucking the blood. The red mites are frequently seen in poultry houses about perches, etc. As they live in cracks and go on to the fowls at night, they can be killed by a free use of kerosene or kerosene emulsion about the perches, etc.

MISCELLANEOUS.

Fowls often die from frosted combs and wattles. Remedies are not so satisfactory as prevention. If fowls are discovered before frost has come out, applications of cold water or snow, till the frost is out, and then an application of vaseline, will sometimes save them.

Fowls become very fond of eating eggs if they once begin, and the habit spreads from fowl to fowl. Great care should be taken to prevent giving them a taste through frozen or broken eggs. An egg-eating fowl, when discovered, should usually be killed.

Feather-eating sometimes becomes a habit like egg-eating, and is most objectionable. The cure is often effected by a change of diet and wide range of food. A few bad cases may be treated by filing the beak so that the back and front will not come together, but not so as to prevent ordinary eating.

Sometimes chickens are poisoned by eating salt meat or fish, or picking up grains of salt. Whites of eggs and liquor of boiled flaxseed are recommended. Laudanum and finely powdered chalk mixed with water, either together or singly, may be given, especially when there appears to be pain.

Danger from rats may usually be obviated if the poultry houses are so constructed that the rats cannot burrow under them. Do not attempt poisoning except in extreme cases. When poison is necessary, use a little strychnine in cheese. The pieces of Cheese should be dropped into the rat holes and covered with boards to prevent fowls from getting them.

The best protection against hawks is covering the yard with wire netting. If there are bushes for the fowls to hide under there will be less danger.

CHAPTER XII.

BELGIAN HARES.

The Belgian Hare differs from all other hares or rabbits in many important particulars. The ears are longer, the eyes larger and more prominent, head broader between the eyes, and they have lighter meat and tougher hide. They are natives of Germany, Belgium, Switzerland and Scotland. They are three times as large as our common rabbits, but very gentle and docile, and thrive best in small quarters. They require but little care, and can be fed on grass, hay, turnips, grain, etc., the same as sheep. They are very prolific, commencing to breed when only five or six months old, producing their young every six or eight weeks, and from six to twelve at a time. Their color is rufus red (red brown), varying somewhat in female to a yellowish or dun cast. Their meat is tender, rich and juicy, and of a fine gamey flavor. A full grown Belgian Hare will weigh from 8 to 12 pounds. In France and Germany their skins are highly valuable for making imitation kid gloves. They are much more profitable than other rabbits and always command a ready sale in the market when game is in season. An enterprising boy or girl can make money by raising Belgian Hares, faster than in any other way. They can be easily kept on what generally goes to waste about a farm, and their management and culture make very pleasant and very profitable employment, either for children or grown people.

Belgian Hare.

Stop a moment to figure; under ordinary circumstances, one doe will produce six litters a year with an average of six youngsters each time; and supposing half of them are does, that would make 18 does in one year. The does of the first litter will also have three litters the first year, which, on the same basis, will give you 27 more. That isn't all. The second litter will have young twice before the first year is up, which will figure 18 more., making a grand total of 63 does the first year from only one with which you started. And yet, that is not all. You will have as many bucks as you have does, which will increase the number to 126 hares from a single dose. Figure the 126 hares to weigh on an average of 8 pounds each, which gives you 1,008 pounds of meat, and at 10 cents a per pound, you will have $100.80. The above average is very low, and is only what anyone can do if he wishes. In raising hares, it is necessary to get the very best stock if good results are to be obtained. The pedigreed Golden Bay Strain is acknowledged to be superior to all others in every respect, and the demand is so large for them that it is almost impossible to secure enough to supply customers.

HOW TO CARE FOR BELGIAN HARES.

They should be fed principally with clover hay, oats and corn. Dry feed is preferable to green food. When grass is fed to hares there is danger of their eating too much, which sometimes proves quite injurious to them. Clover hay is the best food and they are very fond of it. If there is an abundant supply furnished there is little need of anything else. Hares are in many respects about the same as sheep, especially in the food line. They eat anything that a sheep will, cabbage, turnips, plantain leaves, in fact anything in the way of grass and vegetables as well as hay, oats and corn. A box 2 feet high, 3 feet wide and 4 feet long is sufficient room for a doe. Give her plenty of straw or hay, and keep her in nice, clean and comfortable quarters. Provide a box 15 inches wide, 20 inches long and 14 inches high in which she can have her young. This box should have a hole in one side, placed near the top, and should be about five inches in diameter, and should have a lid on top so as to make it easy to clean out. Provide plenty of straw or other good material to make the nest. The doe should be bred two weeks after she has had her young. The young hares when about four weeks old, should be removed to separate quarters. Do not put young hares of different ages in one pen or yard. Keep each lot separate if you wish them to do well. The buck will require about the same room as the doe, excepting he will not need the small box. Do not let their hutches become wet and filthy. Keep them perfectly clean and dry. They may be kept in a barn or any out building convenient for that purpose; or buildings may be especially prepared for them which is the better plan if you wish to raise them in large numbers.

BELGIAN HARE PELTS AND FUR.

Owing to the effort to produce fine specimens for show purposes, but little attention has been paid by the majority of the Belgian hare breeders to the economic value of this little animal, and a few points relative to this side of the industry may be of interest. Owing to the cleanly habits of the hare when properly fed and cared for, the flesh is a table delicacy, rivalling chicken and turkey. The pelt of an animal less than a year old is not sought by the furrier although when properly tanned it is suitable for rugs, capes, etc., and a buggy robe made of the well-selected and well-tanned pelts of animals from five to seven months old, suitably lined, presents an elegant appearance.

The pelts of young animals, however, find a ready market in New York. The fur is removed from the pelt and is used for hat felt, and the pelt is afterwards melted into glue. The price paid for the entire pelt is ten to twenty-five cents per pound, being regulated by the amount and quality of fur on the pelt. Fur of a good quality is worth seventy-five cents per pound. The hare can be raised to five months of age properly and sufficiently fed for three cents per pound actual cost, and as the meat will bring from ten to fifteen cents per pound dressed, it can be readily seen that the balance is on the right side of the ledger.

www.ingramcontent.com/pod-product-compliance
Lightning Source LLC
Chambersburg PA
CBHW031859200326
41597CB00012B/485